Cambridge Paper

A Level Maths Series

Maths Problem Solving using the Casio *fx-991ES* Calculator

Mechanics 2

Dr Allen Brown

Cambridge
Paperbacks

Cambridge Paperbacks

www.CambridgePaperbacks.com

First published by Cambridge Paperbacks 2016

ISBN: 978-0-9935916-8-6

Disclaimer

Although the author and publisher have made every effort to
ensure that the information in this book was correct during
preparation and printing, the author and publisher hereby
disclaim any liability to any party for any errors or omissions.

Read this First

Welcome to the *Part 2* of this two part series on using your *Casio fx-991 ES PLUS* calculator for solving at A Level mechanics problems. At this stage it will be useful to introduce you to a six stage procedure for solving mechanics problems although the procedure is applicable to many other maths problems. When given a maths problem, try to approach it using the following steps.

1. Designate symbols to all numerical quantities given in the problem.

2. Decide which existing general equations to use for the problem.

3. Derive an equation or a set of equations that are particular to the current problem which provide the solutions – a *mathematical model* of the problem.

4. Enter the keystrokes in your *fx-991ES* calculator for the equation(s).

5. Perform the calculations.

6. This is optional: Make slight variations in the input numerical values and observe how they change the solutions.

If you likely to pursue a university degree in either a science, technology, engineering of maths (STEM) subject, you will find this procedure useful as it's similar to model creation and coding.

Allen Brown
Cambridge

Contents

1. Collisions and Impulse

Collisions occupy a special place in the field of mechanics as they are real events that happen every day in one form or another. In the first instant we shall be looking at *elastic* collisions, so no energy is lost and the momentum before the collision is the same as after the collision. Later on we shall be looking at inelastic collisions. In chapter 1 of *Part 1* you were introduced to the concept of the *conservation of momentum*. Also the concept of the *impulse* which is the difference in the momentum of two colliding particles. Impulse is a vector as it has both direction and magnitude. Impulse *I* can be defined as *force × time*, which is equivalent to,

$$I = m(v - u)$$

which is the product of the mass and the change in velocity.

Example 1: A particle has a mass of 5kg and a velocity vector $\begin{pmatrix} 3 \\ 4 \end{pmatrix}$. It is given an impulse of $\begin{pmatrix} -2 \\ 6 \end{pmatrix}$. What is the velocity of the particle after the effect of the impulse, the angle and magnitude of the vector?

Solution: Let the velocity of the particle after the impact be *v*, therefore the impulse becomes,

$$\begin{pmatrix} -2 \\ 6 \end{pmatrix} = 5v - 5 \begin{pmatrix} 3 \\ 4 \end{pmatrix}$$

Rearranging this expression,

$$v = \frac{1}{5} \begin{pmatrix} -2 \\ 6 \end{pmatrix} + \begin{pmatrix} 3 \\ 4 \end{pmatrix} = \frac{1}{5} \begin{pmatrix} -2 \\ 6 \end{pmatrix} + \frac{1}{5} \begin{pmatrix} 15 \\ 20 \end{pmatrix}$$

Leaving

$$v = \frac{1}{5} \begin{pmatrix} -13 \\ 26 \end{pmatrix} \ ms^{-1}$$

To calculate the angle and magnitude, use your *fx* calculator with the following keystrokes,

[SHIFT] (Pol) [(-)] [1] [3] [SHIFT] [(,)] [2] [6] [)] [=]

Giving $|v| = 29.068$ and $\theta = 116.56°$.

Very often you would come across problems where two particles travelling in a straight line and collide and you will expected to calculate any velocity changes that take place.

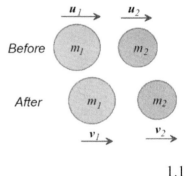

Consider two particles colliding as shown in the diagram on the right. From the conservation of momentum we know that,

$$m_1u_1 + m_2u_2 = m_1v_1 + m_2v_2$$

This may be expressed as,

$$m_1(u_1 - v_1) = m_2(u_2 - v_2)$$

1.1

You will observe from this expression the impulse on m_1 is the same as the impulse on m_2. A frequent question involve the particles sticking together (coalesce) during the collision as shown in the diagram on the right. After the collision they move as a single particle. Therefore,

$$m_1u_1 - m_2u_2 = (m_1+m_2)v$$

The velocity after the collision is therefore

$$v = \frac{m_1u_1 - m_2u_2}{m_1 + m_2}$$

1.2

3

Example 2: Two particles are on a collision course, the first particle is 2 kg with a velocity of 4 ms^{-1} and the second is 3 kg with a velocity of 5 ms^{-1} is moving towards the first particle. During the collision they coalesce into one particle, what is the velocity of the resultant mass?

Solution:
1. Allocate letters to the numerical values given in the problem, $X = 2$ kg, $Y = 3$ kg, $A = 4$ ms^{-1} and $B = 5$ ms^{-1}.
2. Use Eq:1.1 the conservation of momentum equation.
3. Use Eq:1.2 to determine the velocity of the combined mass after the collision.
$$\frac{XA - YB}{X + Y}$$
4. The keystrokes for this expression, (A)

    [▤] [ALPHA] (X) [ALPHA] (A) [−] [ALPHA] (Y) [ALPHA] (B) [▼]
    [RCL] (X) [+] [RCL] (Y) [CALC]

5. After [CALC] has been entered you will be asked for the numerical values, you then see the velocity in the

    ⓪   FIX   Math ▲
    $$\frac{XA-YB}{X+Y}$$

    -1.4000

    display
    The coalesced mass is moving towards the left.
6. At this stage we do not have to make adjustments to the input values.

Example 3: A ball weighing 0.5 kg with velocity vector $\binom{4}{7}$ ms^{-1} is projected onto a wall. The rebound velocity vector was $\binom{2}{3}$ ms^{-1}. Determine the impulse from the wall.

Solution: The momentum before the impact was $0.5\begin{pmatrix}4\\7\end{pmatrix}$ and the momentum after the impact was $0.5\begin{pmatrix}2\\3\end{pmatrix}$ the impulse is therefore,

$$0.5\begin{pmatrix}2\\3\end{pmatrix} - 0.5\begin{pmatrix}4\\7\end{pmatrix} = 0.5\begin{pmatrix}-2\\-4\end{pmatrix} = \begin{pmatrix}-1\\-2\end{pmatrix} \; Ns$$

Example 4: Particle A collides with particle B which is k times more massive than A. The velocity vector of A is $\begin{pmatrix}4\\2\end{pmatrix}$ ms^{-1} and the velocity vector of B is $\begin{pmatrix}6\\-2\end{pmatrix}$ ms^{-1}. After the collision, they coalesce and the resultant velocity vector is $\begin{pmatrix}4\\2\end{pmatrix}$. What is the value of k?

Solution: Momentum before the collision is equal to the momentum after the collision, therefore,

$$\begin{pmatrix}4\\2\end{pmatrix} + k\begin{pmatrix}6\\-2\end{pmatrix} = (k+1)\begin{pmatrix}5.2\\-0.4\end{pmatrix}$$

This becomes,

$$\begin{pmatrix}4\\2\end{pmatrix} + k\begin{pmatrix}6\\-2\end{pmatrix} = k\begin{pmatrix}5.2\\-0.4\end{pmatrix} + \begin{pmatrix}5.2\\-0.4\end{pmatrix}$$

Rearrange

$$k\begin{pmatrix}6\\-2\end{pmatrix} - k\begin{pmatrix}5.2\\-0.4\end{pmatrix} = \begin{pmatrix}5.2\\-0.4\end{pmatrix} - \begin{pmatrix}4\\2\end{pmatrix}$$

This becomes,

$$k\begin{pmatrix}6-5.2\\-2+0.4\end{pmatrix} = \begin{pmatrix}5.2-4\\-0.4+2\end{pmatrix}$$

or

$$k\begin{pmatrix}0.8\\-1.6\end{pmatrix} = \begin{pmatrix}1.2\\-2.4\end{pmatrix}$$

Leaving $k = 1.2/0.8 = 1.5$

Here are a few exercises to work through. You will be expected to calculate the resulting velocity and the coalesced mass?

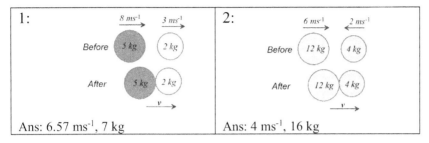

1:

| Before | 8 ms⁻¹ 5 kg | 3 ms⁻¹ 2 kg |

Ans: 6.57 ms⁻¹, 7 kg

2:

Ans: 4 ms⁻¹, 16 kg

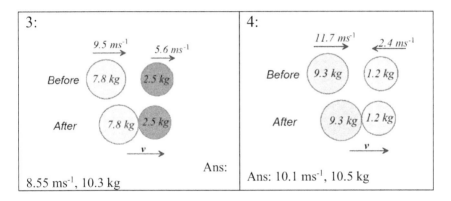

3:

Ans:
8.55 ms⁻¹, 10.3 kg

4:

Ans: 10.1 ms⁻¹, 10.5 kg

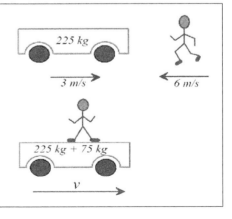

5: A trailer of mass 225 kg is moving with a velocity of 3 ms⁻¹ and a man, weighing 75 kg, is running towards the trailer with a velocity of 6 ms⁻¹ and leaps onto it. What is the new velocity of the trailer with the man standing on it?

Ans:0.75 ms⁻¹.

6: An object of mass 2 kg is subjected to an impulse $\begin{pmatrix} 4 \\ -8 \end{pmatrix} N.$ After the impact the velocity vector is $\begin{pmatrix} 3 \\ 5 \end{pmatrix} ms^{-1}.$ Determine the velocity vector before the impact.

Ans: $\begin{pmatrix} 1 \\ 9 \end{pmatrix} ms^{-1}$

1.1 Inelastic Collisions

Collisions in general are not elastic and a measure of how elastic a collision is, is the *coefficient of restitution* (*e*) first defined by Newton. It's defined as,

$$e = \frac{Relative\ speed\ after\ collision}{Relative\ speed\ before\ collision}$$

This can also be considered as,

$$Speed\ of\ separation = e \times Speed\ of\ approach$$

The range of e is $0 < e < 1$. When $e = 0$, the objects coalesces into a single particle. Whereas when $e = 1$ the collision is perfect elastic. You will probably have come across a *bouncy jet ball*, an example is shown on the right, it has a high value of e close to 1. You could have a collision where $e > 1$, this would be an exploding impact where there would be more kinetic energy after the collision than before. Now consider how this coefficient is used.

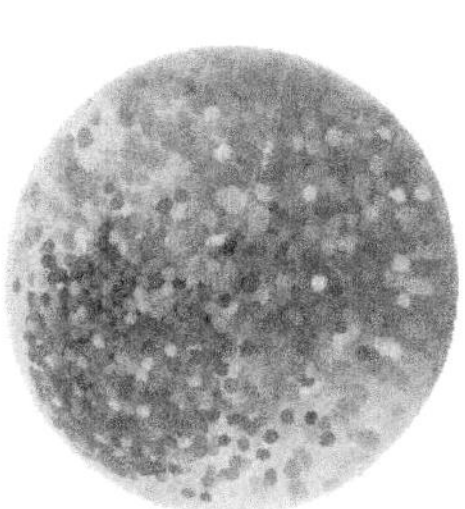

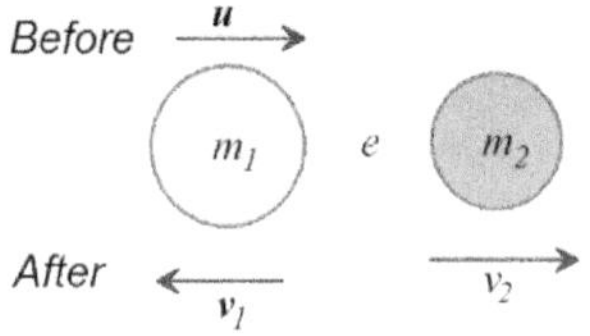

Before $\xrightarrow{u}$

m_1 $\quad e \quad$ m_2

After $\xleftarrow{\quad}$ $\xrightarrow{\quad}$
v_1 $\qquad v_2$

Two particles colliding with one particle at rest as shown in the diagram on the right. By the conservation of momentum,

7

$$m_1 u = m_2 v_2 - m_1 v_1$$

By Newton's Law,

$$e = \frac{v_1 + v_2}{u} \qquad\qquad 1.3$$

The two unknowns are usually v_1 and v_2, these expressions can be rearranged to give,

$$v_1 = \frac{u(em_2 - m_1)}{m_1 + m_2} \qquad\qquad 1.4$$

$$v_2 = \frac{um_1(e+1)}{m_1 + m_2} \qquad\qquad 1.5$$

When calculating v_1 and v_2, you can use Eq:1.4 together with Eq:1.3,

$$v_2 = eu - v_1 \qquad\qquad 1.6$$

When performing this two stage calculation on your *fx* calculator, v_1 is calculated from Eq:1.4 and the [Ans] is fed into Eq:1.6 to give v_2. These expressions can therefore be evaluated on your *fx* calculator using the following *fx* key designations: $X = m_1$, $Y = m_2$, $e = E$ and $A = u$, then,

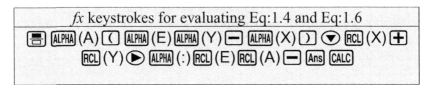

fx keystrokes for evaluating Eq:1.4 and Eq:1.6
[▤] [ALPHA] (A) [(] [ALPHA] (E) [ALPHA] (Y) [−] [ALPHA] (X) [)] [▼] [RCL] (X) [+] [RCL] (Y) [▶] [ALPHA] (:) [RCL] (E) [RCL] (A) [−] [Ans] [CALC]

Example 4: Two balls experience an inelastic collision. The first ball has mass of 1 kg and a velocity 2 ms^{-1} and the second ball has mass 3 kg and is stationary. What are the velocities of the balls after the collision if the value of e is 0.45?

Solution: The diagram shows the two balls before and after the collision.

Before $\xrightarrow{2\ ms^{-1}}$ 1 kg   e   3 kg

After $\xleftarrow{\quad}$ v_1 $\xrightarrow{\quad}$ v_2

1. Allocate symbols to the values, $X = 1$ kg, $Y = 3$ kg, $E = 0.45$ and $A = 2$ ms^{-1}.
2. Use Eq:1.1 and Eq:1.3.
3. Use Eq:1.4 to determine the velocity v_1 and Eq:1.6 to calculate v_2.
4. Use the *fx* keystrokes for evaluating Eq:1.4 and Eq:1.6 (on previous page).
5. After [CALC] has been entered you will be asked for the numerical values, enter these after the last [=] you will see,

FIX Math ▲Disp

$$\frac{A(EY-X)}{X+Y}$$

0.1750

FIX Math ▲

EA-Ans

0.7250

The velocity of first ball is 0.175 ms^{-1} towards the left and the velocity of the second ball is 0.725 ms^{-1}. (Use the [S⇔D] key to change the display format).

6. Varying the input values is performed in the following two exercises.

4-1: Using the same keystrokes as in *Example 4*, what are the velocities after the collision when the mass of the first ball is 4.7 kg with a velocity of 3.4 ms^{-1}. The second ball has a mass of 2.5 kg and is stationary. The value of *e* is 0.67.

Ans: 1.428, 3.706 (both moving in the same direction)

4-2: Again using the same keystrokes as in *Example 4*, what are the velocities after the collision when the mass of the first ball is 1.6 kg with a velocity of 5.7 ms^{-1}. The second ball has a mass of 3.2 kg and is stationary. The value of *e* is 0.53.

Ans: -0.11 ms^{-1}, 2.9 ms^{-1}.

Two particles are approaching each other with masses m_1 and m_2 and velocities v_1 and v_2 as shown in the diagram on the right. Using the conservation of momentum, then

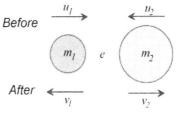

Before

After

$$m_1 u_1 - m_2 u_2 = -m_1 v_1 + m_2 v_2.$$

The value of e is given by,
$$e = \frac{v_1 + v_2}{u_1 + u_2} \qquad\qquad 1.7$$

u_1 and u_2 and known and you need to calculate the values of v_1 and v_2. There are two unknowns, solving these equations we get,

$$v_1 = \frac{u_1(em_2 - m_1) + u_2 m_2(e+1)}{m_1 + m_2} \qquad\qquad 1.8$$

and
$$v_2 = \frac{m_1 u_1(1+e) + u_2(em_1 - m_2)}{m_1 + m_2} \qquad\qquad 1.9$$

As you can see the input variables are m_1, m_2, u_1, u_2 and e. When performing the calculation to determine v_1 and v_2, first calculate v_1 using Eq:1.8, then use,
$$v_2 = e(u_1 + u_2) - v_1 \qquad\qquad 1.10$$

In this two stage calculation on your *fx* calculator, Eq:1.8 provides the $\boxed{\text{Ans}}$ for the second stage to feed into Eq:1.10. Here are the keystrokes for your *fx* calculator with the following designations: $X = m_1$, $Y = m_2$, $e = E$, $A = u_1$, and $B = u_2$.

Keystrokes for Eq:1.8 and Eq:1.10
▦ [ALPHA] (A) [(] [ALPHA] (E) [ALPHA] (Y) [−] [ALPHA] (X) [)] [+] [ALPHA] (B) [RCL] [(]
Y) [(] [RCL] (E) [+] [1] [)] [▼] [RCL] (X) [+] [RCL] (Y) [▶] [ALPHA] (:)
[RCL] (E) [(] [RCL] (A) [+] [RCL] (B) [)] [−] [Ans] [CALC]

10

The following example shows how this set of keystrokes are used to solve a collision problem.

Example 5: Two particles are on a collision course. The first particle has a mass of 3.4 kg and velocity 2.3 ms^{-1} and the second particle has a mass of 4.4 kg and velocity 3.6 ms^{-1}. If the value of e is 0.64, calculate the velocity of the particles after the collision.

Solution: A diagram of the collision is shown on the right, the two unknowns are v_1 and v_2.

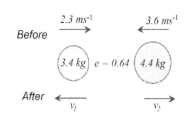

1. Allocate symbols to the values given in the problem, $X = 3.4$ kg, $Y = 4.4$ kg, $E = 0.64$, $A = 2.3$ ms^{-1} and $B = 3.6$ ms^{-1}.
2. Eq:1.1 and Eq:1.7 are the source equations.
3. For this problem Eq:1.8 and Eq:1.10 are used.
4. Use the *fx* keystrokes for evaluating Eq:1.8 and Eq:1.10.
5. After [CALC] has been entered you will be asked for the numerical values, enter these after the last [=] you will see,

$$\frac{A(EY-X)+BY(E+1)}{X+Y}$$

$$3.1583$$

$$E(A+B)-Ans$$

$$0.6177$$

The value of v_1 is 3.158 ms^{-1} to the left for the 3.4 kg mass and the value of v_2 is 0.617 ms^{-1} for the 4.4 kg mass moving to the right.

6. Variations in the input values: if the 4.4 kg mass is increased by 10%, how does this affect the after collision velocities? A 10% increase of 4.4 kg is 4.84 kg. Enter

11

[CALC] and enter 4.84 for *Y*. This gives, $v_1 = 3.38$ ms^{-1} and $v_2 = 0.39$ ms^{-1}.

5-1: Using the same keystrokes as *Example 5*; the first particle has a mass of 4.5 kg and velocity 3.9 ms^{-1} and the second particle has a mass of 6.3 kg and velocity 8.7 ms^{-1}. If the value of *e* is 0.63, calculate the velocity of the particles after the collision.

Ans: 7.05 ms^{-1}, -0.632 ms^{-1} (moving towards the left)

Example 6: A ball is dropped from a height of 2 m and weighs 0.25 kg. The coefficient of restitution between the ball and the ground is 0.7. Calculate the height reached after the first bounce and the impulse between the ball and the ground.

Solution:
1. Let $X = 0.25$ kg, $D = 2$ m and $E = 0.7$.
2. The equation of motion to use is $v^2 = u^2 + 2as$.
3. The initial velocity is zero and the velocity of the ball as it touches the ground is
$$v = \sqrt{2gD} \qquad a$$
The rebound velocity is given by,
$$u = Ev = E\sqrt{2gD} \qquad b$$
The height reached on the rebound is calculated from the equation of motion,
$$height = \frac{u^2}{2g} \qquad c$$
The impulse *I* is given by,
$$I = m(v + u) \qquad d$$
The solution involves four calculations *a, b, c* and *d*. (A)
4. The *fx* allocation, $A = u$, $B = v$, $X = 0.25$ and $D = 2$m, the keystrokes are,

12

a

ALPHA (B) ALPHA (=) $\sqrt{\blacksquare}$ 2 SHIFT (CONST) 3 5 ALPHA (D) ▶ ALPHA (:)

b ALPHA (A) ALPHA (=) ALPHA (E) RCL (B) ALPHA (:)

c RCL (A) x^2 🖶 2 SHIFT (CONST) 3 5 ▶ ALPHA (:)

d ALPHA (X) ⟮ RCL (B) ⊟ RCL (A) ⟯ CALC

    5.   Perform the calculation

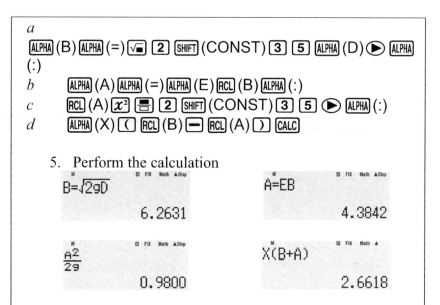

$$B=\sqrt{2gD}$$
$$\text{6.2631}$$

$$A=EB$$
$$\text{4.3842}$$

$$\frac{A^2}{2g}$$
$$\text{0.9800}$$

$$X(B+A)$$
$$\text{2.6618}$$

The velocity of impact is 6.26 ms⁻¹, the velocity of rebound is 4.38 ms⁻¹, the height of the rebound is 0.98 m and the impulse is 2.66 *Ns*.

    6.   What effect will there be if the coefficient of restitution is reduced to 0.62? The rebound velocity is 3.88 ms⁻¹, the rebound height is 0.768 m and the impulse is 2.54 *Ns*.

6-1: Using the same *fx* keystrokes as in *Example 6*, calculate the impact velocity, the rebound velocity, the height of the rebound and the impulse when the ball is dropped from a height of 3.6 m and the coefficient of restitution is 0.81.

Ans: 8.4 ms⁻¹, 6.91 ms⁻¹, 2.36 m and 3.8 *Ns*.

In the following exercises calculate the values of the v_1 and v_2.

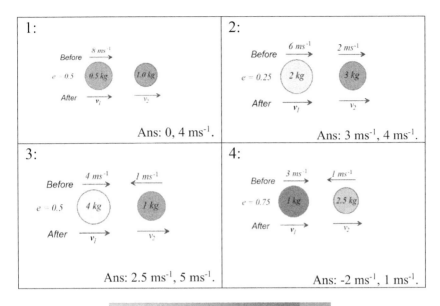

1:

Before $\xrightarrow{8\ ms^{-1}}$

$e = 0.5$ 0.5 kg   1.0 kg

After $\xrightarrow{v_1}$ $\xrightarrow{v_2}$

Ans: 0, 4 ms⁻¹.

2:

Before $\xrightarrow{6\ ms^{-1}}$ $\xrightarrow{2\ ms^{-1}}$

$e = 0.25$ 2 kg   3 kg

After $\xrightarrow{v_1}$ $\xrightarrow{v_2}$

Ans: 3 ms⁻¹, 4 ms⁻¹.

3:

Before $\xrightarrow{4\ ms^{-1}}$ $\xleftarrow{1\ ms^{-1}}$

$e = 0.5$ 4 kg   1 kg

After $\xrightarrow{v_1}$ $\xrightarrow{v_2}$

Ans: 2.5 ms⁻¹, 5 ms⁻¹.

4:

Before $\xrightarrow{3\ ms^{-1}}$ $\xleftarrow{1\ ms^{-1}}$

$e = 0.75$ 1 kg   2.5 kg

After $\xrightarrow{v_1}$ $\xrightarrow{v_2}$

Ans: -2 ms⁻¹, 1 ms⁻¹.

Cambridge Paperbacks
A Level Maths Series

**Problem Solving
using the Casio
fx-991ES Calculator**

Allen Brown

Mechanics Part 1

2. Work, Energy and Power

Generally mechanics is all about movement and the energy involved in that movement. Whenever an object is moving it has kinetic energy given by,

$$KE = \frac{1}{2}mv^2 \qquad\qquad 2.1$$

where m is the mass of the object and v is it's velocity. When looking at mechanical systems, energy is changed from one form to another. In chapter 1 of Part 1 you will recall that energy comes in several forms, kinetic, potential, heat, nuclear, solar, geothermal, elastic, electrical and chemical to name just a few. In this chapter we shall be looking at kinetic and potential energy which is related to the height an object is above the ground. Whenever an interaction takes place energy will be exchanged (converted into a different form). The unit of energy is the joule (J). So if an object of mass 5 kg is moving with a velocity of 3 ms^{-1}, its kinetic energy, according to Eq:3.1, is $0.5 \times 5 \times 3^2 = 22.5$ J.

2.1 Collisions and Kinetic Energy

A favourite topic for questions on energy usually involves collisions. In reality when a collision takes place, energy will be converted from kinetic energy into heat. Consider the collision shown in the diagram on the right between mass m_1 and mass m_2. The initial velocities are u_1 and u_2 and the velocities after the collision are v_1 and v_2. By the conservation of momentum,

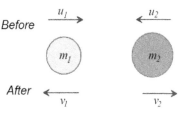

$$u_1 m_1 - u_2 m_2 = -m_1 v_1 + m_2 v_2 \qquad\qquad 2.2$$

Also if it was a perfect elastic collision, by the conservation of energy,

$$\frac{1}{2}m_1u_1^2 + \frac{1}{2}m_2u_2^2 = \frac{1}{2}m_1v_1^2 + \frac{1}{2}m_2v_2^2 \qquad 2.3$$

However for an inelastic collision, the loss in kinetic energy is,

$$KE_{loss} = \frac{1}{2}[m_1(u_1^2 - v_1^2) + m_2(u_2^2 - v_2^2)] \qquad 2.4$$

As you can appreciated in this equation there are six variables. You will probably come across problems where after the impact one of the masses will remain at rest. Let m_2 remain at rest after the collision, $v_2 = 0$. The momentum equation becomes,

$$u_1m_1 - u_2m_2 = -m_1v_1 \qquad 2.5$$

And the energy loss equation becomes,

$$KE_{loss} = \frac{1}{2}[m_1(u_1^2 - v_1^2) + m_2u_2^2] \qquad 2.6$$

From Eq:2.5,

$$v_1 = \frac{u_2m_2}{m_1} - u_1 \qquad 2.7$$

This equation can be used with Eq:2.6 to calculate the energy loss. Here are the fx-991ES keystrokes for this calculation where $X = m_1$, $Y = m_2$, $A = u_1$ and $B = u_2$,

fx keystrokes for Eq:2.7 and Eq:2.6
[ALPHA] (B) [ALPHA] (Y) [▤] [ALPHA] (X) [▶] [−] [ALPHA] (A) [ALPHA] (:) [0] [·] [5] [(] [RCL] (X) [(] [RCL] (A) [x²] [−] [Ans] [x²] [)] [+] [RCL] (Y) [RCL] (B) [x²] [)] [CALC]

The following example shows how these keystrokes are used,

Example 1: Mass G of 2 kg and has velocity of 4 ms^{-1} is approaching mass H of 3 kg also with a velocity of 4 ms^{-1} travelling in the opposite direction. After the collision mass H remains stationary. What is the velocity of mass G after the collision and the loss in kinetic energy?

16

Solution:

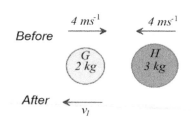

1. Allocate symbols to the values given in the problem, $X = 2$ kg, $Y = 3$ kg, $A = 4$ ms^{-1} and $B = 4$ ms^{-1}.

2. Determine which equation are to be used, Eq:2.2 and Eq: 2.4.

3. These are modified to Eq:2.7 and Eq:2.6.

4. Use the fx keystrokes for Eq:2.7 and Eq:2.6 given on the previous page.

5. After [CALC] has been entered you will be asked for the numerical values, enter these after the last [=] you will see,

$$\frac{BY}{X} - A$$

⓪ FIX Math ▲Disp

2.0000

$$0.5(X(A^2 - Ans^2) \triangleleft \triangleright$$

⓪ FIX Math ▲

36.0000

The velocity of G is 2 ms^{-1} and the loss in kinetic energy is 36 J

1-1: Using the keystrokes as in *Example 1*, mass G has a weight of 3.6 kg and a velocity of 4.7 ms^{-1}. Mass H has a weight of 5.4 kg and a velocity of 6.3 ms^{-1}. After the collision H is stationary, calculate the velocity of G and the loss of kinetic energy in the collision.

Ans: 4.75 ms^{-1}, 106.3 J

2.2 Coalescing Collisions

A variation on this theme is when the two masses, after they collide, they combine (coalesce) to form a single mass as illustrated in the diagram on the next page. Mass G collides with mass H to form a single mass J. The combined mass J is,

$$m_3 = m_1 + m_2$$

The momentum equation is,

$$u_1 m_1 - u_2 m_2 = -v_1 m_3$$

The velocity after the collision is,

$$v_1 = \frac{u_2 m_2 - u_1 m_1}{m_3} \qquad 2.8$$

The difference in the kinetic energy is,

$$KE_{loss} = \frac{1}{2}m_1 u_1^2 + \frac{1}{2}m_2 u_2^2 - \frac{1}{2}m_3 v_1^2 \qquad 2.9$$

To perform these calculations on you fx calculate, let $X = m_1$, $Y = m_2$, $M = m_3$, $A = u_1$ and $B = u_2$.

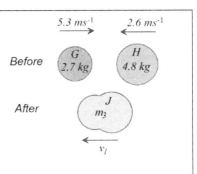

fx Keystrokes for Eq:2.8 and Eq:2.9
[🖶] [ALPHA] (B) [ALPHA] (Y) [−] [ALPHA] (A) [ALPHA] (X) [▼] [RCL] (M) [▶] [ALPHA] (:)
[0] [·] [5] [(] [RCL] (X) [RCL] (A) [x^2] [+] [RCL] (Y) [RCL] (B) [x^2] [−] [RCL]
(M) [Ans] [x^2] [)] [CALC]

Before these keystrokes are used, you will need to store the value of $m_1 + m_2$ in memory M. The following example will demonstrate how these keystrokes are used.

Example 2: Mass G of weight 2.7kg has a velocity of 5.3 ms^{-1}. It is travelling towards Mass H which has a weight of 4.8 kg and a velocity of 2.6 ms^{-1}. During the collision they combine to form Mass J. Calculate the velocity of J after the collision and how much kinetic energy is lost.

Solution: Use the keystrokes for Eq:2.8 and Eq:2.9. First perform the calculation for $m_3 = m_1 + m_2$.

$$\boxed{2}\ \boxed{\cdot}\ \boxed{7}\ \boxed{+}\ \boxed{4}\ \boxed{\cdot}\ \boxed{8}\ \boxed{\text{SHIFT}}\ (\text{STO})(\text{M})\ \boxed{\text{AC}}$$

Now enter the keystrokes with the *fx* designations, $u_2 = B = 2.6$ ms^{-1}, $m_2 = Y = 4.8$ kg, $u_1 = A = 5.3$ ms^{-1} and $m_1 = X = 2.7$ kg.

M 田 FIX Math ▲Disp	M 田 FIX Math ▲
BY−AX	$0.5(XA^2+YB^2-MAr$ ▶
M	
−0.2440	53.9222

The velocity of m_3 is 0.244 ms^{-1} moving from left to right and the loss in kinetic energy is 53.92 J.

2-1: Using the keystrokes as in *Example 2*, $u_1 = 8$ ms^{-1}, $u_2 = 5$ ms^{-1}, $m_1 = 6$ kg and $m_2 = 4$ kg. Calculate the velocity of the coalesced mass and the loss in kinetic energy. (When the display show M? enter 10, sum of the masses).

Ans: 2.8 ms^{-1} and 202.8 J

2.3 Forces at Work

Whenever a force moves it involves energy. This energy is called *work*. Consider the equation of motion,

$$v^2 = u^2 + 2as$$

Multiply by ½*m*,

$$\frac{1}{2}mv^2 = \frac{1}{2}mu^2 + ams$$

Since Force = mass × acceleration, then

$$Fs = \frac{1}{2}mv^2 - \frac{1}{2}mu^2 \qquad\qquad 2.10$$

This indicated that a change in kinetic energy (the right hand side of the equation) is equal to force × distance. This is the *work done* by the force acting over a distance *s*.

Example 3: The mass of a car is 1,300 kg, its velocity of 10 ms^{-1} before it begins to brake. It comes to a halt in 23 m. Calculate (a) the work done in bringing the car to a halt, (b) the force of the braking.

Solution: (a) the work done in bringing the car to a halt is the same as the KE of the vehicle,

$$KE = \frac{1}{2}mv^2 = \frac{1}{2} \times 1300 \times 10^2 = 65\ kJ$$

(b) From Eq:2.10

$$F = \frac{1}{2s}mv^2 = \frac{1300 \times 10^2}{2 \times 23} = 2,826\ N$$

The fx keystrokes for the general problem, where the fx designations are $m = M$, $v = X$ and $s = D$,

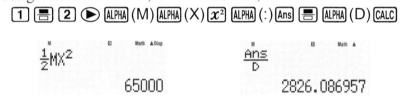

$$\frac{1}{2}MX^2$$

$$65000$$

$$\frac{Ans}{D}$$

$$2826.086957$$

3-1: Using the same keystrokes as *Example 3*, calculate the work needed to bring a car to a halt when its weight is 1,450 kg travelling at 23 ms^{-1} and stopping in a distance of 42 m.

Ans: 383.5 kJ, 9,131.5 N.

Example 4: A car of weight 1,300 kg is travelling along a level road and coefficient of friction between the road and the

tires is 0.61. If the engine is producing a f orce of 11,500 N, what is acceleration of the car and what is the energy used after 250 m?

20

Solution: There are two forces acting on the car, its engine and the friction. The reaction of the road to the car is $R = mg$ and the frictional force is μmg. The total force acting on the car is therefore $F - \mu mg$. Since force = mass × acceleration, the acceleration of the car is,

$$a = \frac{F-\mu mg}{m} = \frac{11500-0.61\times1300\times9.81}{1300} = 2.86 \; ms^{-2}$$

The work done is force × distance, therefore,

$$W_{done} = dma$$

where d is the distance travelled. Here are the keystrokes for your *fx-991* calculator, with the designation, M = 1300 kg, E = 0.61, F = 11,500 N and D = 250 m,

[ALPHA] (A) [ALPHA] (=) [⬛] [ALPHA] (F) [−] [ALPHA]
(E) [ALPHA] (M) [SHIFT] (CONST) [3] [5] [▼] [RCL] (M) [▶] [ALPHA] (:) [ALPHA]
(D) [RCL] (M) [Ans] [CALC]

M	⊞	Math ▲

$A = \dfrac{F-EM9}{M}$

2.864097346

M	⊞	Math ▲

DMAns

930831.6375

The acceleration is 2.86 ms^{-2} and the work done is 930 kJ

4-1: Using the same keystrokes as in *Example 4*, calculate the acceleration of the car and the work done when the engine force is 9,500 N, the mass is increased by 91 kg and the distance is 72 m.

Ans: 0.847 ms^{-2}, 84.88 kJ

2.4 Work and Potential Energy of Objects on Inclined Planes

Another source of problems is the movement of objects on inclined planes. For example an object as shown in the diagram on the right. If h is the height of the incline, the distance of the incline is.

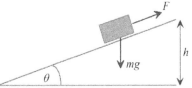

$$L = \frac{h}{\sin(\theta)}$$

The reaction of the incline to the object is $mg \cos(\theta)$ and the frictional force is $\mu mg \cos(\theta)$. The work done by the weight moving downwards is

$$Work_{done} = mgL \sin(\theta) - \mu mgL \cos(\theta)$$

Expressing this in terms of height h,

$$Work_{done} = mgh - \mu mgh \cot(\theta)$$

or

$$Work_{done} = mgh[1 - \mu \cot(\theta)]$$

The $Work_{done}$ must be equal to the kinetic energy, therefore,

$$\frac{1}{2}mv^2 = mgh[1 - \mu \cot(\theta)]$$

Leaving,

$$v = \sqrt{2gh[1 - \mu \cot(\theta)]} \qquad\qquad 2.11$$

This is the velocity gained by the object as it slides along the full length of the incline. The kinetic energy of the object as it reaches the base of the incline will be,

$$KE = \frac{1}{2}mv^2 \qquad\qquad 2.12$$

The fx keystrokes for calculating v and KE are, using the following fx designations; $h = D$, $\theta = F$, $E = \mu$, and $m = M$.

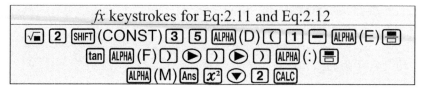

fx keystrokes for Eq:2.11 and Eq:2.12
$\sqrt{\blacksquare}$ [2] [SHIFT] (CONST) [3] [5] [ALPHA] (D) [(] [1] [−] [ALPHA] (E) [🖶]
[tan] [ALPHA] (F) [)] [▶] [)] [▶] [)] [ALPHA] (:) [🖶]
[ALPHA] (M) [Ans] [x^2] [▼] [2] [CALC]

Example 5: An object of mass 5.2 kg is at the top of an inclined plane of height 3.5 m. If the inclination of the plane is 26° and the coefficient of friction is 0.48, what is the velocity and the kinetic energy of the object when it reaches the base of the plane?

Solution: Use the keystrokes for Eq:2.11 and 2.12, with $M = $ 5.3 kg, $D = 3.5$ m, $F = 26°$ and $E = 0.48$,

$$\sqrt{2gD\left(1 - \frac{E}{\tan(F)}\right)}$$

$$1.043232053$$

$$\frac{MAns^2}{2}$$

$$2.884082756$$

The final velocity is 1.04 ms^{-1} and the kinetic energy is 2.88 J

5-1: Using the same keystrokes as in Example 4, what is the final velocity and kinetic energy when object mass is 7.8 kg, the height of the plane is 10.8 m, the inclination angle is 31° and the coefficient of friction is 0.22?

Ans: 11.59 ms^{-1}, 523.6 J

When an object is at height h, it potential energy is given by,

$$PE = mgh \qquad\qquad 2.13$$

23

where *m* is its mass. Normally for a falling object, its penitential energy is converted into kinetic energy. However when an object is sliding down a rough incline, some of the potential energy is converted into heat. The percentage energy loss due to friction is given by,

$$E_{loss} = \left(1 - \frac{KE}{PE}\right) 100 \qquad\qquad 2.14$$

This calculation can be incorporated into the *fx* keystrokes for Eq:2.11 and Eq:2.12.

fx keystrokes for Eq:2.11, Eq:2.12 and Eq:2.14
[√■] [2] [SHIFT] (CONST) [3] [5] [ALPHA] (D) [(] [1] [−] [ALPHA] (E) [⬚] [tan] [ALPHA] (F) [)] [▶] [)] [▶] [)] [ALPHA] (:) [⬚] [ALPHA] (M) [Ans] [x²] [▼] [2] [▶] [ALPHA] (:) [(] [1] [−] [⬚] [Ans] [▼] [SHIFT] (CONST) [3] [5] [RCL] (M) [RCL] (D) [▶] [)] [×] [1] [0] [0] [CALC]

Example 6: An object of mass 7.1 kg is at the top of an inclined plane of height 7.4 m. If the inclination of the plane is 34° and the coefficient of friction is 0.38, what is the velocity, the kinetic energy of the object when it reaches the base of the plane and the percentage loss of energy?

Solution: Use the keystrokes for Eq:2.11, 2.12 and Eq:2.14, with $M = 7.1$ kg, $D = 7.4$ m, $F = 34°$ and $E = 0.38$,

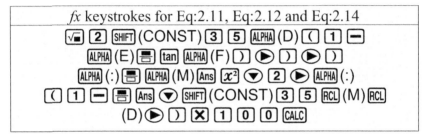

$$\sqrt{2gD\left(1 - \frac{E}{\tan(F)}\right)} \qquad \frac{M \, Ans^2}{2} \qquad \left(1 - \frac{Ans}{gMD}\right) \times 100$$

$$7.960611065 \qquad\qquad 224.9682162 \qquad\qquad 56.3373168$$

The final velocity is 7.96 ms⁻¹, the kinetic energy is 224.97 J and the percentage of energy loss is 56.34%.

A variation on this problem is a car travelling up an incline as illustrated in the diagram on the right. The car's engine has to provide enough force to overcome the effect of gravity and the frictional force of the 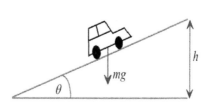 road. The gravity force parallel to the plane is $mg \sin(\theta)$. and will act in the same direction as the frictional force which is $\mu mg \cos(\theta)$. The length of the incline L is related to its height by,

$$h = L \sin(\theta)$$

The work done to overcome the friction is

$$Work_{friction} = \mu mgL \cos(\theta)$$

The work done to overcome the gravity,

$$Work_{gravity} = mgL \sin(\theta)$$

If the force from the engine is F, the work done by the engine is,

$$Work_{engine} = FL$$

If the car starts from the base of the incline with velocity v_0 and the velocity at the top of the incline is v_1, the work energy equation is,

$$\frac{1}{2}mv_1^2 = \frac{1}{2}mv_0^2 + Work_{engine} - Work_{friction} - Work_{gravity}$$

25

This can be rewritten as,

$$\tfrac{1}{2}mv_1^2 = \tfrac{1}{2}mv_0^2 + FL - \mu mgL\cos(\theta) - mgL\sin(\theta)$$

or

$$KE_{final} = \tfrac{1}{2}mv_0^2 + L\{F - mg[\mu\cos(\theta) + \sin(\theta)]\} \quad 2.15$$

The car's velocity at the top of the incline is,

$$v_1 = \sqrt{\frac{2KE_{final}}{m}} \qquad 2.16$$

The keystrokes for your *fx* calculator to perform the calculations in Eq:2.15 and Eq:2.16 with the *fx* designations $L = D$, $m = M$, $F = F$, $\mu = A$, $X = v_0$ and $E = \theta$.

fx keystrokes for Eq:2.15 & Eq:2.16
[ALPHA] (M) [ALPHA] (X) $[x^2]$ [▤] [⊙] [2] [▶] [+] [ALPHA] (D) [(] [ALPHA] (F) [−]
[RCL] (M) [SHIFT] (CONST) [3] [5] [(] [ALPHA] (A) [cos] [ALPHA] (E) [)] [+]
[sin] [RCL] (E) [)] [)] [)] [ALPHA] (:) [√■] [2] [Ans] [▤] [RCL] (M) [CALC]

Example 7: A car weighing 1250 kg begins the ascent of a hill with a velocity of 12ms⁻¹. The angle of the slope is 26°, the car has travelled 43 m up the slope, the coefficient of friction is 0.6 and its engine produces a force of 11,000 *N*.

Solution: Use the keystrokes for Eq:2.15 & Eq:2.16 with *fx* designated values, $D = 43$, $M = 1250$ kg, $F = 12{,}000$ *N*, $A = 0.6$, $X = 12$ ms⁻¹ and $E = 26°$.

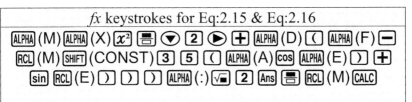

```
M              ▦    Math ▲Disp            M           ▦    Math ▲
MX²                                       ╱2Ans
─── +D(F-Mg(Acos⯈                         ╲╱ ───
 2                                            M
        90674.69219                              12.0448955
```

The kinetic energy is 90.67 kJ and the velocity after 43 m is 12 ms⁻¹.

7-1: Using the same keystrokes as in *Example 6*, what is the kinetic energy and final velocity when a car weighs 1300 kg, whose engine provides a force of 13,000 N and has an initial velocity of 13.7 ms⁻¹ and driven 52 m up the same slope?

Ans: 149.88 kJ, 15.185 ms⁻¹.

On a similar theme, a particle of mass m is projected up an inclined rough plane with an initial velocity u as shown in the diagram on the right. Typical calculations include:

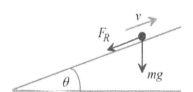

1. The total force F acting on the particle.

2. The acceleration of the particle (this will be negative).

3. The work done after a distance s.

4. The velocity at distance s alone the plane.

5. The kinetic energy at distance s.

6. Distance travelled along the plane before it stops.

There will be two forces acting on the motion of the particle, gravity and friction. The total force is therefore,

$$F = mg \sin(\theta) + \mu mg \cos(\theta)$$

or

$$F = mg[\sin(\theta) + \mu \cos(\theta)] \qquad 2.17$$

Since force = mass × acceleration, the acceleration is given by,

$$a = \frac{F}{m} \qquad\qquad 2.18$$

The work done by the particle (force × distance) as it travels up the slope is,

$$Work_{done} = mas \qquad\qquad 2.19$$

The velocity at distance s is given by,

$$v = \sqrt{u^2 - 2as} \qquad\qquad 2.20$$

The kinetic energy of the particle when its velocity is v is given my,

$$KE = \frac{1}{2}mv^2 \qquad\qquad 2.21$$

The maximum distance the particle travels up the plane is derived from,

$$v^2 = u^2 - 2as$$

$v = 0$ at the maximum distance,

$$S_{max} = \frac{u^2}{2a} \qquad\qquad 2.22$$

Given these six equations, here are the *fx-991* keystrokes for performing all of these calculations, with the *fx* letter designations, $u = X$, $m = M$, $\mu = F$, $s = D$ and $\theta = E$.

fx Keystrokes for Eq:2.17 – Eq:2.22	
$F = mg[\sin(\theta)$ $+ \mu\cos(\theta)]$	[ALPHA] (M) [SHIFT] (CONST) [3] [5] [(] [sin] [ALPHA] (E) [)] [+] [ALPHA] (F) [cos] [RCL] (E) [)] [)] [ALPHA] (:)
$a = \frac{F}{m}$	[ALPHA] (A) [ALPHA] (=) [▤] [Ans] [▼] [RCL] (M) [▶] [ALPHA] (:)
$Work_{done} = mas$	[RCL] (M) [Ans] [ALPHA] (D) [ALPHA] (:)

28

$v = \sqrt{u^2 - 2as}$	[√■] [ALPHA] (X) $[x^2]$ [−] [2] [RCL] (A) [RCL] (D) [▶] [ALPHA] $(:)$
$KE = \frac{1}{2}mv^2$	[0] [•] [5] [RCL] (M) [Ans] $[x^2]$ [ALPHA] $(:)$
$S_{max} = \dfrac{u^2}{2a}$	[▤] [RCL] (X) $[x^2]$ [▼] [2] [RCL] (A) [CALC]

Example 8: A particle of weight 2.1 kg is launched up an inclined slope of 25° with an initial velocity of 14.4 ms⁻¹. If the coefficient of friction is 0.32, calculate (a) the force acting on the particle, (b) the acceleration of the particle, (c) the work done after it has travelled 3.6 m, (d) its velocity at 3.6 m, (e) the kinetic energy at 3.6 m and (f) the maximum distance it will travel up the slope.

Solution: Use all the *fx* keystrokes for Eq:2.17 – Eq:2.22 with the *fx* letter designations, $X = 14.4\ ms^{-1}$, $M = 2.1\ kg$, $F = 0.32$, $D = 3.6\ m$ and $E = 25^o$.

Mg(sin(E)+Fcos(▷	A=$\frac{\text{Ans}}{\text{M}}$	MAnsD
14.67601636	6.98857922	52.8336589

$\sqrt{X^2-2AD}$	0.5MAns²	$\frac{X^2}{2A}$
12.53164912	164.8943411	14.8356335

The downward force acting of the particle is 14.67 *N*, the acceleration is -6.98 ms⁻², the work done after 3.6 m is 52.83 *J*, the velocity at this distance is 12.53 ms⁻¹, the kinetic energy of the particle is 164.89 *J* and the particle stops after travelling 14.8 m.

8-1: Using the same keystrokes as in *Example 6*: a particle of mass 5 kg is projected up a rough inclined slope whose angle is 23° with a launch speed of 5.7 ms^{-1}. If the coefficient of friction is 0.33, calculate:
(a) the force acting on the particle,
(b) the acceleration of the particle,
(c) the work done after it has travelled 2.1 m,
(d) its velocity at 2.1 m,
(e) the kinetic energy at 2.1 m
(f) the maximum distance it will travel up the slope.

Ans: 34 N, 6.8 ms^{-2}, 71.5 J, 1.97 ms^{-1}, 9.71 J, 2.3 m

8-2: Again using the same keystrokes as in *Example 6*: a particle of mass 4.3 kg is projected up a rough inclined slope whose angle is 31° with a launch speed of 6.4 ms^{-1}. If the coefficient of friction is 0.32, calculate:
(a) the force acting on the particle,
(b) the acceleration of the particle,
(c) the work done after it has travelled 1.9 m,
(d) its velocity at 1.9 m,
(e) the kinetic energy at 1.9 m
(f) the maximum distance it will travel up the slope.

Ans: 33.2 N, 7.74 ms^{-2}, 63.2 J, 3.4 ms^{-1}, 24.8 J, 2.64 m

2.5 Power

Power is a measure of the rate energy is used and is measured in Watts. One Watt of power occurs when one Joule of energy is used in one second. As you know,

$$Work_{done} = Force \times distance$$

Therefore

$$Power = \frac{Work_{done}}{time} = Force \times \frac{distance}{time} = Force \times velocity.$$

When dealing with electrical appliances the electrical power they produce is the product of the voltage and the current flowing through them. Generally, the hotter an electrical appliance, the more power it dissipates.

Very often power calculations involve accelerating cars. In the diagram on the right, the car has a forward force of F, a resistive force F_R due to the road resistance. Its current velocity is v, its acceleration is a and its weight is m. If the engine power is P and the car is travelling with velocity v, the force F generated by the engine is,

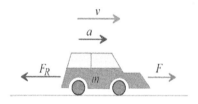

$$F = \frac{P}{v} \qquad\qquad 2.23$$

and the acceleration is given by,

$$a = \frac{F - F_R}{m} \qquad\qquad 2.24$$

When the car is not accelerating, the force generated by the engine equals the resistive force F_R and the velocity of the car is,

$$v = \frac{P}{F_R} \qquad\qquad 2.25$$

This is effectively the maximum velocity of the car with a fixed amount of power. The keystrokes for your *fx* calculator, when you are given the power P, the current velocity v, the mass m and the resistive force F_R, with the *fx* designations: $X = P$, $v = Y$, $A = a$, $M = m$ and $F_R = F$,

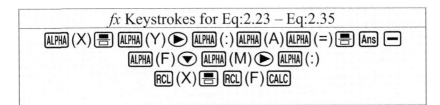

	fx Keystrokes for Eq:2.23 – Eq:2.35	

[ALPHA] (X) [▤] [ALPHA] (Y) [▶] [ALPHA] (:) [ALPHA] (A) [ALPHA] (=) [▤] [Ans] [−]
[ALPHA] (F) [▼] [ALPHA] (M) [▶] [ALPHA] (:)
[RCL] (X) [▤] [RCL] (F) [CALC]

Example 9: A car weighing 1,000 kg is travelling and is meeting a resistive force of 475 N. The car's engine is producing 4 kW of power. What is the force produced by the engine, the acceleration when its velocity is 5 ms^{-1} and the maximum velocity of the car?

Solution: Using the *fx* keystrokes for Eq:2.23 – Eq:2.35, with $X = 4,000$ W, $Y = 5$ ms^{-1}, $M = 1,000$ kg and $F = 475$ N.

$\frac{X}{Y}$	$A=\frac{Ans-F}{M}$	$\frac{X}{F}$
800	0.325	8.421052632

The force from the engine is 800 *N*, the acceleration when the velocity is 5 ms^{-1} is 0.325 ms^{-2} and the maximum speed for the 4kW engine is 8.41 ms^{-1}.

9-1: Using the same keystrokes as *Example 7*, the car weighing 1,500 kg is travelling and is meeting a resistive force of 397 N. The car's engine is producing 5.5 kW of power. What is the force produced by the engine, the acceleration when its velocity is 6.1 ms^{-1} and the maximum velocity of the car?

Ans: 901.6 N, 0.336 ms^{-2}, 13.85 ms^{-1}

When a car is travelling up an incline extra demands are placed on the engine to provide the power. In the diagram on the right, the car has mass m, and the resistive force is F_R and the force generated by the engine is F.

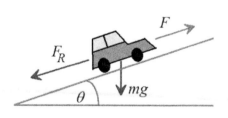

The angle of the incline is θ. If the velocity of the car is v, then F is related to the power by,

$$F = \frac{P}{v} \qquad\qquad 2.26$$

The car has to produce a force to overcome gravity, the resistive effect is therefore given by,

$$F_R = F - mg\sin(\theta) \qquad\qquad 2.27$$

Once the car levels out it will accelerated where,

$$ma = F - F_R$$

The acceleration is therefore given by

$$a = \frac{F - F_R}{m} \qquad\qquad 2.28$$

The fx keystrokes for your calculator to perform the calculations in Eq:2.26 – Eq:2.28 when given the mass, the power and the velocity of the car with the fx designations: $X = P$, $v = Y$, $E = \theta$ and $M = m$,

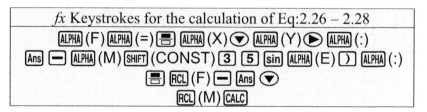

Example 10: A car of weight 1,200 kg is climbing an incline whose angle is 3.8° at a velocity of 25 ms⁻¹. The engine power is 25 kW. Calculate the force generated by the engine, total resistance to the car's motion and the acceleration once the car is on the level.

Solution: Use the *fx* keystrokes for Eq:2.26 – 2.28 with $X = 2,500$ W, $M = 1,200$ kg and $E = 3.8°$.

$$F = \frac{X}{V}$$

1250

Ans–Mgsin(E)

465.9913444

$$\frac{F - Ans}{M}$$

0.6533405464

The force generated by the engine is 1,250 *N*, the total resistance to the car's motion is 466 *N* and the acceleration on the level road is 0.653 ms⁻².

10-1: Using the same keystrokes as *Example 8*, if the car weighed 1,250 kg, was driving up an incline of 5° with a velocity of 37 ms⁻¹ and the engine power was 40 kW. What is the force from the engine, the total resistive force on the car and its acceleration when it levels off?

Ans: 1081.1 N, 12.7 N, 0.85 ms⁻².

Consideration is often given to the coupling between a vehicle and a trailer, in particular the tension T the coupling bar 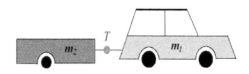 which has to sustain the tension as the vehicle is moving. The figure on the right shows the coupling bar between car and trailer with horizontal tension T. The weight of the vehicle is m_1 and the weight of the trailer is m_2. Let the resistive force of the trailer be

34

F_t and the resistive force of the car be F_v . If F is the driving force of the vehicle which not only causes the vehicle to move but also the trailer. Then we have the net force from the vehicle is,

$$F - F_v - T = m_1 a \qquad 2.29$$

As far as the trailer is concerned, it will be propelled to the right by the tension T in the coupling therefore the net force on the trailer is,

$$T - F_t = m_2 a \qquad 2.30$$

where a is the acceleration. When we add Eq:2.29 and Eq:30 the acceleration becomes,

$$a = \frac{F - (F_v + F_t)}{m_1 + m_2} \qquad 2.31$$

Example 11: A small car weighing 150 kg is connected to trailer with a weight of 52 kg. The force generated in the engine is 500 N and the resistive force on the car is 12 N and on the trailer 6.8 N. Calculate the acceleration of the car and the tension in the coupling between them.

Solution: To calculate the acceleration, use Eq:2.31 and to calculate the tension use either Eq:2.29 or Eq:2.30. When performing this calculation on your *fc-991ES* calculator, it will be in two stages. The first to calculate the acceleration and the second to calculate the tension using Eq:2.29.
Use the letter designations; $X = m_1 = 150$ kg, $Y = m_2 = 52$, $F = 500$ N, $A = F_v = 12$ N, and $B = F_t = 6.8$ N. The keystrokes are,

⊟ [ALPHA] (F) ⊟ [(] [ALPHA] (A) ⊞ [ALPHA] (B) [)] ▼ [ALPHA] (X) ⊞
[ALPHA] (Y) ▶ [ALPHA] (:) [RCL] (F) ⊟ [RCL] (A) ⊟ [RCL] (X) [Ans] (X) [CALC]

$$\frac{F-(A+B)}{X+Y}$$

2.382178218

F-A-XAns

130.6732673

The acceleration is 2.38 ms^{-2} and the tension in the coupling is 130.67 N.

11-1: A locomotive weighing 10,000 kg is connected to a carriage weighing 3,000 kg. The Resistive force on the locomotive is 1,000 N and on the carriage 500 N. If the force generated by the locomotive is 50,000 N what is the acceleration and the tension in the coupling?

Ans: 3.73 ms^{-2}, 11,692.3 N

Example 12: A car weighing 1,700 kg is coupled to a trailer weighing 920 kg. The resistive force on the car is 900 N and on the trailer 330 N. If the car engine is generating a power of 36 kW and its current velocity is 21 ms^{-1}, what is the force produced by the engine, the acceleration of the car and the tension in the coupling?

Solution: Using Eq:2.26, the force from the engine can be expressed as using *fx-991ES* letter designation,

$$F = \frac{power}{velocity} = \frac{E}{C}$$

The acceleration is

$$a = \frac{F-(A+B)}{X+Y}$$

and the tension is given by,

$$T = Ya + B$$

The keystrokes for your *fx-991ES* calculator with, $E = 36,000$ W, $C = 21$ ms^{-1}, $A = 920$ N, $B = 330$ N, $X = 1,700$ kg and $Y = 920$ kg, are

[ALPHA] (F) [ALPHA] $(=)$ [▤] [ALPHA] (E) [▼] [ALPHA] (C) [▶] [ALPHA] $(:)$ [RCL] (F) [━]
[$($] [ALPHA] (A) [+] [ALPHA] (B) [$)$] [▼] [ALPHA] (X) [+] [ALPHA] (Y) [▶] [ALPHA] $(:)$
[RCL] (Y) [Ans] [+] [RCL] (B) [CALC]

E? [3] [6] [0] [0] [0] [=]
C? [2] [1] [=]
A? [9] [2] [0] [=]
B? [3] [3] [0] [=]
X? [1] [7] [0] [0] [=]
Y? [9] [2] [0] [=] [S↔D]

$F = \dfrac{E}{C}$
1714.285714

$\dfrac{F-(A+B)}{X+Y}$
0.1772082879

$Y\text{Ans}+B$
493.0316249

The force produced by the engine is 1,712.28 N, the acceleration is 0.177 ms^{-2} and the tension in the coupling is 493.03 N.

When a vehicle is ascending an inclined plane and drawing a trailer, the effects of gravity acting on both vehicle and trailer needs to be considered. The diagram on the right shows a vehicle going up an

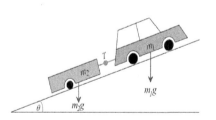

inclined plane. The current velocity, the power generated by the engine and the force from the engine is given by Eq:2.26. From Eq:2.29,

$$F - F_v - T - m_1 g\ \sin(\theta) = m_1 a \qquad 2.32$$

and from Eq:2.30,

$$T - F_t - m_2 g \sin(\theta) = m_2 a \qquad 2.33$$

Adding these equations and rearranging,

$$a = \frac{F-(F_v+F_t)-(m_1+m_2)g\sin(\theta)}{m_1+m_2}$$ 2.34

Also rearranging Eq:2.33 for the tension in the coupling,

$$T = F_t + m_2[g\sin(\theta) + a]$$ 2.35

The *fx-991ES* can be used to perform these calculations, first express Eq:26, Eq:2.34 and Eq:35 using *fx-991ES* letter designations; for the force produced by the vehicle,

$$F = \frac{E}{C}$$ 2.36

The acceleration is,

$$a = \frac{F - (A + B) - (X + Y)g\sin(D)}{X + Y}$$ 2.37

and the tension in the coupling is.

$$T = B + Y[g\sin(\theta) + a]$$ 2.38

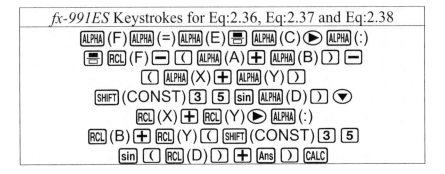

---
fx-991ES Keystrokes for Eq:2.36, Eq:2.37 and Eq:2.38

ALPHA (F) ALPHA (=) ALPHA (E) 冒 ALPHA (C) ▶ ALPHA (:)
冒 RCL (F) ⊟ 〔 ALPHA (A) ⊞ ALPHA (B) 〕 ⊟
〔 ALPHA (X) ⊞ ALPHA (Y) 〕
SHIFT (CONST) ③ ⑤ sin ALPHA (D) 〕 ▼
RCL (X) ⊞ RCL (Y) ▶ ALPHA (:)
RCL (B) ⊞ RCL (Y) 〔 SHIFT (CONST) ③ ⑤
sin 〔 RCL (D) 〕 ⊞ Ans 〕 CALC

Example 13: A car is drawing a trailer up a gradient of angle θ whose value is 8.6° as shown in the diagram on the right. The weight of the car is 1,400 kg and the weight of the trailer is 400 kg. The car engine is working at 60 kW, the resistance on the car is 800 N and on the trailer 300 N. When the speed of the car is 12 ms^{-1}, calculate the magnitude of the car's acceleration and the tension in the coupling connecting the car and trailer.

Solution: Use the above keystrokes to perform the calculations with the following values; $E = 60$ kW, $C = 12$ ms^{-1}, 800 N, $B = 300$ N, $X = 1,400$ kg, $Y = 400$ kg, $D = 8.6°$.

$$F = \frac{E}{C}$$
5000

$$\frac{F-(A+B)-(X+Y)gs}{X+Y}$$
0.7002258909

$$B+Y(gsin(D)+Ans)$$
1166.666667

The force produced by the car is 5,000 N, the acceleration is 0.7 ms^{-2} and the tension in the coupling is 1,166.6 N.

13-1: Using the same keystrokes as *Example 11*, where the angle is 9.3°, the weight of the car is 1,560 kg travelling at 15ms^{-1}, the weight of the trailer is 480 kg, the resistance on the car is 760 N, the resistance on the trailer is 320 N and the engine is working at 73 kW. Calculate the force produced by the car, the acceleration and the tension in the coupling.

Ans: 4.8866 N, 0.271 ms^{-2}, 1,211 N

1: A car is travelling up a hill at whose inclination angle is 3.8°. The velocity is constant at 20 ms^{-1} and the weight of the car is 1,200 kg. If the power generated by the engine is 25 kW. Calculate the force from the engine, the resistance to the car's movement and the acceleration when the ground levels out.

Ans: 1250 N, 466 N, 0.953 ms^{-2}.

2.6 Variable Resistive Forces

In the examples so far, an assumption has been made that the forces on the vehicle are constant.

We can now include forces that are a function of the vehicle's velocity using the equation,

$$ma = F - f(v)$$

Replacing F with the power to velocity ratio,

$$m\frac{dv}{dt} = \frac{P}{v} - f(v) \qquad\qquad 2.39$$

This equation can be expressed in two ways, the first for a time calculation and the second for a distance calculation. For a time calculation Eq:2.39 can be written as,

$$\int_{v_1}^{v_2} \frac{mv}{P - vf(v)}\,dv = \int_0^T 1\,dt$$

Leaving

$$T = m\int_{v_1}^{v_2} \frac{v}{P - vf(v)}\,dv \qquad\qquad 2.40$$

T is the time taken for the vehicle to accelerate from v_1 to v_2 with a fixed power P generated by the engine. For a distance calculation, the acceleration is written as,

$$\frac{dv}{dt} = \frac{dv}{dy}\frac{dy}{dt} = v\frac{dv}{dy}$$

Eq:2.29 becomes,

$$mv\frac{dv}{dy} = \frac{P}{v} - f(v)$$

This can be rewritten as,

40

$$Y = m \int_{v_1}^{v_2} \frac{v^2}{P - vf(v)} dv \qquad\qquad 2.41$$

where Y is the distance travelled by the vehicle as it accelerates from v_1 to v_2.

The following example illustrates how Eq:2.40 and Eq:2.41 can be used.

Example 14: The weight of a car is 560 kg and the resistive forces acting against is motion are $f(v) = 80v$ where v is it's current velocity. The engine is generating a power of 72 kW. How long does it take to accelerate from 10 ms^{-1} to 20 ms^{-1} and what is the distance covered.

Solution: Using Eq:2.40, the integral can be performed, however we shall use the definite integration feature on our *fx-991ES* calculator. Expressing the integral use *fx-991ES* letter designations,

$$560 \int_{10}^{20} \frac{X}{72,000 - 80X^2} dX$$

The keystrokes for this calculation are,

[5] [6] [0] [∫☐] [▤] [ALPHA] [(X)] [⌄] [7] [.] [2] [x10ˣ] [4] [−] [8] [0]
[RCL] [(X)] [x²] [⌄] [1] [0] [▲] [2] [0] [=]

$$560 \int_{10}^{20} \frac{X}{7.2 \times 10 4 - 80}$$
$$1.645012702$$

The time taken to accelerate from 10 ms^{-1} to 20 ms^{-1} is 1.64 seconds. To calculate the distance, use Eq:2.41

$$560 \int_{10}^{20} \frac{X^2}{72,000 - 80X^2} dX$$

The keystrokes for this calculation are,

$$\boxed{\text{AC}}\,\boxed{5}\,\boxed{6}\,\boxed{0}\,\boxed{\int_\square^\square}\,\boxed{\equiv}\,\boxed{\text{ALPHA}}\,(\text{X})\,\boxed{x^2}\,\boxed{\blacktriangledown}\,\boxed{7}\,\boxed{\cdot}\,\boxed{2}\,\boxed{\times 10^x}\,\boxed{4}\,\boxed{-}$$
$$\boxed{8}\,\boxed{0}\,\boxed{\text{RCL}}\,(\text{X})\,\boxed{x^2}\,\boxed{\blacktriangledown}\,\boxed{1}\,\boxed{0}\,\boxed{\blacktriangle}\,\boxed{2}\,\boxed{0}\,\boxed{=}$$

$$560 \int_{10}^{20} \frac{x^2}{7.2 \times 10^4 - 80} \triangleright$$
$$26.21052685$$

The distance taken to achieve this acceleration is 26.21 m.

Example 15: The weight of a car is 620 kg and the resistive forces acting against is motion are $f(v) = 2.3v^2$ where v is it's current velocity. The engine is generating a power of 81 kW. How long does it take to accelerate from 10 ms^{-1} to 22 ms^{-1} and what is the distance covered.

Solution: Using Eq:2.40 with the *fx-991ES* letter designations,
$$620 \int_{10}^{22} \frac{X}{81,000 - 2.3X^3} dX$$

The *fx-991ES* keystrokes for this calculation are,
$$\boxed{\text{AC}}\,\boxed{6}\,\boxed{2}\,\boxed{0}\,\boxed{\int_\square^\square}\,\boxed{\equiv}\,\boxed{\text{ALPHA}}\,(\text{X})\,\boxed{\blacktriangledown}\,\boxed{8}\,\boxed{\cdot}\,\boxed{1}\,\boxed{\times 10^x}\,\boxed{4}\,\boxed{-}\,\boxed{2}$$
$$\boxed{\cdot}\,\boxed{3}\,\boxed{\text{RCL}}\,(\text{X})\,\boxed{\text{SHIFT}}\,(x^3)\,\boxed{\blacktriangledown}\,\boxed{1}\,\boxed{0}\,\boxed{\blacktriangle}\,\boxed{2}\,\boxed{2}\,\boxed{=}$$

$$620 \int_{10}^{22} \frac{X}{8.1 \times 10^4 - 2.} \triangleright$$
$$1.743823737$$

The time taken to accelerate from 10 ms^{-1} to 22 ms^{-1} is 1.74 seconds. The distance covered in this time is calculated by using Eq:2.41,
$$620 \int_{10}^{22} \frac{X^2}{81,000 - 2.3X^3} dX$$

The *fx-991ES* keystrokes for this calculation are,

[AC] [6] [2] [0] [∫▢] [▢] [ALPHA] (X) [x²] [▼] [8] [.] [1] [×10ˣ] [4] [−]
[2] [.] [3] [ALPHA] (X) [SHIFT] (x³) [▼] [1] [0] [▲] [2] [2] [=]

$$620 \int_{108.1 \times 104-2.}^{22} \frac{x^2}{} \triangleright$$

29.76293109

The distance travelled in 1.74 seconds is 29.76 m.

1: The weight of a car is 750 kg and the resistive forces acting against is motion are $f(v) = 90 \ln(2v)$ where v is its current velocity. The engine is generating a power of 77 kW. How long does it take to accelerate from 12 ms^{-1} to 32 ms^{-1} and what is the distance covered.

$$750 \int_{12}^{32} \frac{X}{77,000 - 90X \ln(2X)} \, dX$$

$$750 \int_{12}^{32} \frac{X^2}{77,000 - 90X \ln(2X)} \, dX$$

Ans: 4.8 s, 112.87 m

2: The weight of a car is 850 kg and the resistive forces acting against is motion are $f(v) = 0.2 \, e^{0.3v}$ where v is its current velocity. The engine is generating a power of 63 kW. How long does it take to accelerate from 9 ms^{-1} to 21 ms^{-1} and what is the distance covered.

$$850 \int_{9}^{21} \frac{X}{63,000 - 0.2Xe^{0.3X}} \, dX$$

$$850 \int_{9}^{21} \frac{X^2}{63,000 - 0.2Xe^{0.3X}} \, dX$$

Ans: 2.45 s, 38.85 m

3: The weight of a car is 870 kg and the resistive forces acting against is motion are $f(v) = \ln(1 + v^2)$ where v is its current velocity. The engine is generating a power of 86 kW. How long does it take to accelerate from 15 ms^{-1} to 28 ms^{-1} and what is the distance covered.

$$870 \int_{15}^{28} \frac{X}{86,000 - X \ln(1 + X^2)} \, dX$$

$$870 \int_{15}^{28} \frac{X^2}{86,000 - X \ln(1 + X^2)} \, dX$$

Ans: 2.83 s, 62.74 m

4: The weight of a car is 902 kg and the resistive forces acting against is motion are $f(v) = 67\sqrt{v}$ were v is its current velocity. The engine is generating a power of 97 kW. How long does it take to accelerate from 12 ms^{-1} to 32 ms^{-1} and what is the distance covered.

$$902 \int_{12}^{32} \frac{X}{97,000 - 67X^{1.5}} \, dX$$

$$902 \int_{12}^{32} \frac{X^2}{97,000 - 67X^{1.5}} \, dX$$

Ans: 4.45s, 105.45 m

3. Statics and Limited Equilibrium

Whenever an object is subjected to forces it will usually move. On occasions the forces cancel each other out resulting in the object being stationary – in a state of equilibrium. When the object in is just about to move, this is called *limited equilibrium*. A frequent source of problems are rods leaning against walls on the point of slippage, the following example illustrates this point.

Example 1:

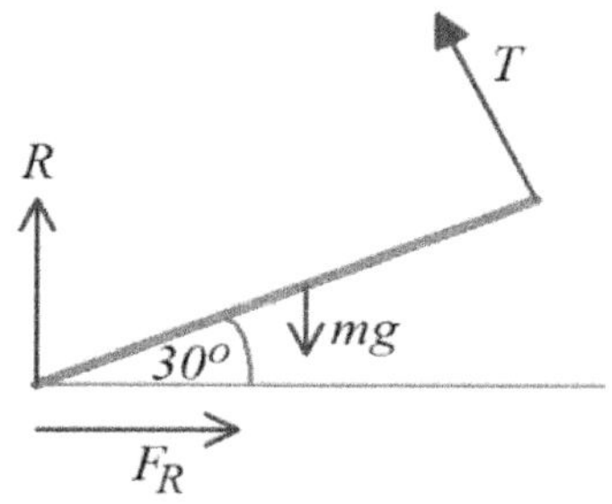

- A uniform rod weighing 10 kg is leaning against a wall as seen in the diagram on the right.

- One end of the rod rests on a rough surface while the other end is supported by a perpendicular rope whose tension is T.

- The angle of the rod is 30° and the frictional force that stops the rod from slipping is F_R.

Calculate the following: the tension T in the rope, the magnitude of R, the frictional force F_R and smallest possible value for the coefficient of friction to sustain the equilibrium.

Solution:

1. Allocate symbols to values, $M = 10$ kg, $C = 30°$,
2. The horizontal forces are
$$F_R - T\sin(C) = 0 \qquad\qquad 1$$
The vertical forces are
$$R + T\cos(C) - mg = 0 \qquad\qquad 2$$
If a is the length of the rod, the moment about the point touching the rough surface is,

---

45

$$mg\frac{a}{2}\cos(C) - Ta = 0 \qquad\qquad 3$$

The reaction R is related to the frictional force F by,
$$F_R = \mu R \qquad\qquad 4$$

3. Rearrange these three equations, the tension is given from 3,
$$T = \frac{mg\cos(C)}{2} \qquad\qquad 5$$

The frictional force F is given from 1,
$$F_R = T\sin(C) \qquad\qquad 6$$

The reaction R is given from 2,
$$R = mg - T\cos(C) \qquad\qquad 7$$

The coefficient of friction μ from 4,
$$\mu = \frac{F_R}{R} \qquad\qquad 8$$

4. Need to calculate Eq:5 to Eq:8. The *fx* symbols allocation, let $A = T$ and $F = F_R$, the keystrokes are,

5
[ALPHA] (A) [ALPHA] (=) [▤] [ALPHA] (M) [SHIFT] (CONST) [3] [5] [cos] [ALPHA] (C)
[)] [▼] [2] [▶] [ALPHA] (:)
6   [ALPHA] (F) [ALPHA] (=) [Ans] [×] [sin] [RCL] (C) [)] [ALPHA] (:)
7   [RCL] (M) [SHIFT] (CONST) [3] [5] [−] [RCL] (A)
[×] [cos] [(] [RCL] (C) [)] [ALPHA] (:)
8   [▤] [RCL] (F) [▼] [Ans]

5. Perform the calculation [CALC] with $M = 10$ kg and $C = 30^o$.

46

The tension in the rope is 42.46 *N*, the magnitude of *R* is 61.292 *N*, the frictional force is 21.23 *N* and the minimum coefficient of friction to maintain equilibrium is 0.346.

6. What effect does increasing the angle to 38° have on the equilibrium? The tension is 38.63 *N*, the frictional force is 23.78 *N*, the reaction is 67.61 *N* and the coefficient of friction needs to be 0.352.

1-1: Using the same keystrokes as *Example 1*, calculate the tension in the rope, the frictional force, the reaction and the minimum coefficient of friction when the rod weighs 13.4 kg and the inclination angle is 27°.

Ans: 58.54 N, 26.57 N, 79.24 N, 0.335

3.1 Ladders leaning against walls

Another source of problems are the static forces involved to maintain a ladder of length *L* leaning against a wall as observed in the diagram on the right. There are a number of forces acting on the ladder,

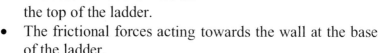

- Its weight mg.
- The reaction of the ground on its base.
- The reaction of the wall on the top of the ladder.
- The frictional forces acting towards the wall at the base of the ladder.

Since the ladder is in equilibrium, the forces cancel out,

horizontal forces: $F_R = B$

47

vertical forces: $R = mg$.

Take moments about the base of the ladder (clockwise forces will cancel anticlockwise forces),

$$mg \cos(C)\frac{L}{2} = B \sin(C)L$$

Rearranging,

$$B = \frac{mg}{2\tan(C)} \qquad\qquad 3.1$$

If μ is the coefficient of friction between the ladder base and the ground, then $F_R = \mu R = \mu mg = B$. The coefficient of friction is therefore,

$$\mu = \frac{B}{mg} \qquad\qquad 3.2$$

These two equations can be calculated on your *fx-991ES* calculator.

fx-991ES keystrokes for Eq:3.1 and Eq:3.2
[ALPHA] (B) [ALPHA] (=) [⊟] [ALPHA] (M) [SHIFT] (CONST) [3] [5] [▼] [2] [tan] [ALPHA] (C) [)] [▶] [ALPHA] (:) [⊟] [Ans] [▼] [RCL] (M) [SHIFT] (CONST) [3] [5] [CALC]

Example 2: A ladder of mass 10 kg rests against a smooth vertical wall. When ladder is inclined at an angle of 60°, calculate the reaction of the wall and the coefficient of friction.

Solution: Using the keystrokes for Eq:3.1 and Eq:3.2 with $M = $ 10 kg and angle $C = 60°$. The display will show,

M ⊡ Math ▲Disp	M ⊡ Math ▲
$B=\frac{M9}{2\tan(C)}$	Ans M9
28.30936009	0.2886751346

The reaction is 28.31 *N* and the coefficient of friction is 0.29

You may have to consider a boy walking up a ladder as observed in the diagram on the right. There will be a distance when the ladder slips – the equilibrium ceases. Again let the length of the ladder be L and the distance reached by the boy is X before it slips (*ignoring Health and Safety*). There will be a frictional force F_R initially preventing the ladder from slipping. R is the reaction of the rough surface to the base of the ladder and B is the reaction of the wall to the top of the ladder. Resolving the forces,

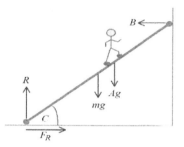

horizontal forces: $F_R = B$
vertical forces: $R = mg + Ag$

where A is the mass of the man. But $F_R = \mu R$ therefore,
$$F_R = \mu R = \mu g(m + A)$$
or
$$B = \mu g(m + A) \qquad\qquad 3.3$$

Taking moments around the base of the ladder,
$$BL \sin(C) = \frac{mg \cos(C)L}{2} + Ag \cos(C) X$$

Rearranging this equation,
$$X = \frac{L}{A}\left[\frac{B}{g}\tan(C) - \frac{m}{2}\right] \qquad\qquad 3.4$$

49

Let $D = L$ and the coefficient of friction $= E$, the *fx* keystrokes are,

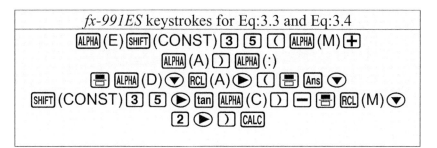

fx-991ES keystrokes for Eq:3.3 and Eq:3.4
[ALPHA] (E) [SHIFT] (CONST) [3] [5] [(] [ALPHA] (M) [+]
[ALPHA] (A) [)] [ALPHA] (:)
[=] [ALPHA] (D) [▼] [RCL] (A) [▶] [(] [=] [Ans] [▼]
[SHIFT] (CONST) [3] [5] [▶] [tan] [ALPHA] (C) [)] [−] [=] [RCL] (M) [▼]
[2] [▶] [)] [CALC]

Example 3: A ladder of length 5 m, weighting 30 kg is leaning against a smooth wall with an inclined angle of 63°. The coefficient of friction between the base of the ladder and the ground is 0.3. A boy weighing 30 kg ascends the ladder. What distance does he reach before it slips?

Solution: Use the keystrokes for Eq:3.3 and Eq:3.4, where M = 30 kg, A = 30 kg, C = 63°, E = 0.3 and D = 5 m. The display will show,

$$E9(M+A)$$
$$176.5197$$

$$\frac{D}{A}\left(\frac{Ans}{9}\tan(C)-\frac{M}{2}\right)$$
$$3.387831517$$

The reaction of the wall is 176.5 *N* and the distance climbed before the ladder slips is 3.39 m.

3-1: Using the same keystrokes as *Example 3*; the ladder is 4 m and its weight is 22 kg. If the weight of the boy is 35 kg and the inclination angle is 50° what distance will he reach before it slips if the coefficient of friction is 0.4?

Ans: 1.85 m

3-2: Referring to *Example 3*, to ensure the ladder does not slip, prove the minimum inclination angle is given by the following expression and calculate that angle,

$$C_{min} = \tan^{-1}\left[\frac{g}{B}\left(A + \frac{m}{2}\right)\right]$$

Ans: 68.2°

Cambridge Paperbacks
A Level Maths Series

Problem Solving
using the Casio
fx-991ES Calculator

Allen Brown

Calculus I

4. Moment of Inertia

A definition of inertia is given in the following box,

> **Inertia** *n* **1.** inertness, slowness to take action. **2.** the property of matter by which it remains in a state of rest or if it's in motion, continues moving in a straight line, unless acted upon by an external force.

It's a means to describe the reluctance of an object to move. The moment of inertia arises when an object is in a state of rotation. Normally the rotation will occur around the *centre of gravity* also referred to as the *centre of mass*. Most objects will have more than one moment of inertia depending on which axis it is rotated through. In our discussion we shall only consider one at most. A simple definition of moment of inertia of an object is

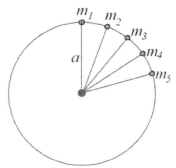

$$\text{mass} \times \text{distance}^2 .$$

Different objects will have different moments of inertia and a simple example is a uniform hoop which is *considered to be circle with a large number of little weights that are closely packed and uniformly distributed over its circumference* as shown in the figure on the right – a hoop with radius a. The moment of inertia for this hoop is defined as,

$$I = a^2 m_1 + a^2 m_2 + a^2 m_3 + a^2 m_4 + \cdots a^2 m_n$$

or

$$I = a^2(m_1 + m_2 + m_3 + m_4 + \cdots m_n)$$

52

where m_n is the nth little weight making up the hoop. The weight of the hoop is therefore the sum of all the little weights which is M, therefore,

$$I = Ma^2 \qquad\qquad 4.1$$

4.1 Moment of Inertia of a Uniform Rod

The moment of inertia of a solid rod about the central axis can be determined by considering it in small sections. The following diagram illustrates the point.

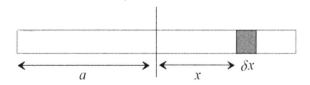

The rod is length $2a$ and mass m, the mass per unit length is $\frac{m}{2a}$. We now consider a small element distance x from the centre of mass, the thickness of the element is δx. The moment of inertia for this single element is mass × length^2, or,

$$\delta I = \left(\frac{m}{2a}\delta x\right)x^2$$

To obtain the moment of inertia for the whole rode this expression needs to be integrated with respect to x,

$$I = \int_{-a}^{a}\frac{m}{2a}x^2 dx = \frac{m}{2a}\int_{-a}^{a}x^2 dx$$

$$\frac{m}{2a}\left[\frac{x^3}{3}\right]_{-a}^{a} = \frac{m}{6a}[a^3 - (-a)^3]$$

Leaving

$$I = \frac{ma^2}{3} \qquad\qquad 4.2$$

This is the moment of inertia of a uniform rod of mass m and length $2a$.

4.2 Moment of Inertia of a Uniform Disc

When considering the moment of inertia of a uniform disc it's necessary to model it as many closely packed annular rings.

The diagram on the right shows a single annulus of thickness δx at a distance x from the centre of the disc. We need to calculate the area of the annulus, this is given by,

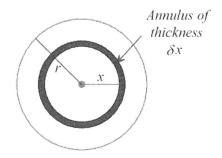

Annulus of thickness δx

$$\delta A = \pi(x + \delta x)^2 - \pi x^2$$

Since $(\delta x)^2$ is small it can be ignored, therefore,

$$\delta A = 2\pi x \, \delta x$$

If the disc is of mass m, the mass per unit area is $\frac{m}{\pi r^2}$. The mass of the annulus is therefore,

$$2\pi x \, \delta x \left(\frac{m}{\pi r^2}\right) = \frac{2mx}{r^2} \, \delta x$$

The moment of inertia for this annulus is therefore,

$$\delta I = \left(\frac{2mx}{r^2} \, \delta x\right) x^2$$

The moment of inertia for the whole disc is given by integrating this expression between the limits of 0 and r, to give,

$$I = \int_0^r \frac{2m}{r^2} x^3 \, dx = \frac{2m}{r^2} \int_0^r x^3 \, dx = \frac{1}{2} mr^2 \qquad\qquad 4.3$$

4.3 Moment of Inertia of a Solid Sphere

When determining the moment of inertia of a sphere, it necessary to consider a sphere made of closely packed discs. In the figure on the right, a disc section is shown and a distance x from the centre of the sphere. The radius of the disc is $\sqrt{r^2 - x^2}$ and if the thickness of the disc is δx, the volume is

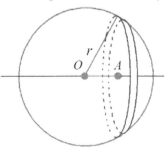

$$\delta V = \pi(r^2 - x^2) \, \delta x$$

The mass per unit volume of the sphere is,

$$\frac{m}{\frac{4}{3}\pi r^3} = \frac{3m}{4\pi r^3}$$

The mass of the disc is therefore,

$$\pi(r^2 - x^2) \, \delta x \frac{3m}{4\pi r^3} = \frac{3m}{4r^3}(r^2 - x^2) \, \delta x$$

The moment of inertia of the disc is given by Eq:4.3,

$$\delta I = \frac{1}{2} \frac{3m}{4r^3}(r^2 - x^2) \, \delta x \, (r^2 - x^2) = \frac{3m}{8r^3}(r^2 - x^2)^2 \, \delta x$$

The moment of inertia for the whole sphere is the integral of this expression between the limits of $-r$ to r,

$$I = \frac{3m}{8r^3} \int_{-r}^{r} (r^2 - x^2)^2 \, dx$$

55

If you have already covered calculus, you should be able to perform this integration by expanding the bracket to give the moment of inertia for a solid sphere is,

$$I = \frac{2}{5}mr^2$$

4.4 Moment of Inertia of a Solid Cylinder

The moment of inertia of a solid cylinder along its axis is calculated in a manner similar to the disc.

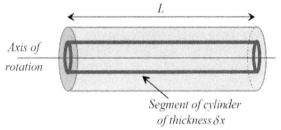

L

Axis of
rotation

Segment of cylinder
of thickness δx

A cylinder is made up of closely packed cylindrical segments as illustrated in the above diagram. If the mass of the cylinder is m and its volume is $\pi r^2 L$ where L is its length, the mass per unit volume is $\frac{m}{\pi r^2 L}$. The volume of the cylindrical section of radius x is its perimeter times its thickness times its length which is $2\pi x \, \delta x$ L. The moment of inertia of this segment is therefore,

$$\delta I = \left(\frac{m}{\pi r^2 L}\right) 2\pi x \, \delta x \, L \, x^2 = \frac{2mL}{r^2} x^3 \, \delta x$$

The moment of inertia for the whole cylinder is this expression integrated between 0 and r,

$$I = \frac{2m}{r^2} \int_0^r x^3 \, dx$$

Leaving,

$$I = \frac{1}{2}mr^2 \qquad\qquad 4.4$$

which is the same as that for a disc.

56

4.5 Moment of Inertia of a Hollow Cylinder

Instead of a cylinder being solid, we can consider the case where a section has been removed from it as see in the figure below,

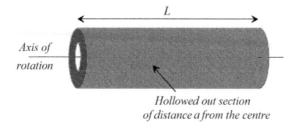

Hollowed out section
of distance a from the centre

The outer radius is b and inner radius is a. The volume of the cylinder is the difference between the whole cylinder and the section removed, which is,

$$\pi b^2 L - \pi a^2 L = \pi L(b^2 - a^2)$$

If the mass is m, the mass per unit volume is,

$$\frac{m}{\pi L(b^2 - a^2)}$$

Using the same argument as in the previous section, the moment of inertia for a section of the remaining cylinder is,

$$\delta I = \left(\frac{m}{\pi L(b^2 - a^2)}\right) 2\pi x \, \delta x \, L \, x^2$$

Now integrate this expression between the limits a and b,

$$I = \int_a^b \left(\frac{2m}{b^2 - a^2}\right) x^3 \, dx = \frac{2m}{b^2 - a^2} \int_a^b x^3 \, dx$$

Perform the integration,

$$I = \frac{2m}{b^2 - a^2} \left[\frac{x^4}{4}\right]_a^b = \frac{m}{2(b^2 - a^2)} [b^4 - a^4]$$

This can be expressed as, using the difference of squares,

$$I = \frac{m}{2(b^2-a^2)}[(b^2-a^2)(b^2+a^2)]$$

The resulting expression for the moment of inertia of a hollow cylinder is therefore,

$$I = \frac{1}{2}m(b^2+a^2) \qquad\qquad 4.5$$

Example 1: Calculate the moment of inertia of a hollow cylinder with an od of 40.6 cm and an id of 20.8 cm and of mass 3.6 kg.

Solution: Use Eq:4.5 where M = 3.6, B = 0.406 m and A = 0.208 m, the *fx* keystrokes for the calculation are,

$$\boxed{1}\;\boxed{\blacksquare}\;\boxed{2}\;\boxed{\blacktriangleright}\;\boxed{\text{ALPHA}}\;(M)\;\boxed{(}\;\boxed{\text{ALPHA}}\;(B)\;\boxed{x^2}\;\boxed{+}\;\boxed{\text{ALPHA}}\;(A)\;\boxed{x^2}\;\boxed{)}\;\boxed{\text{CALC}}$$

$$\tfrac{1}{2}M(B^2+A^2)$$

$$0.3746$$

The moment of inertia is 0.3746 *kg m²*.

4.6 The Additive Rule

Very often objects are made up of combinations of standard shapes and the moment of inertia of these *compound objects* can be determined by adding their separate moments of inertia. To illustrate this point, consider the object shown in the figure below.

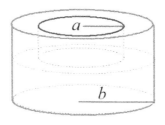

This is a compound structure made up of a solid cylinder base of radius b (whose mass is m_D) and an annulus (of mass m_A) whose internal radius is a. We have already calculated the moment of inertia for a cylinder in *Section 4.4* and also the moment of inertia for a hollow cylinder in *Section 4.5*, by the *Additive Rule* the moment of inertia for the combination as shown in the above figure is,

$$I = \frac{1}{2}m_D b^2 + \frac{1}{2}m_A(b^2 + a^2)$$

This can be written as,

$$I = \frac{1}{2}[b^2(m_D + m_A) + m_A a^2] \qquad\qquad 4.6$$

Example 2: A compound object comprises a base cylinder weighing 5.3 kg and of diameter 5.9 cm and an annulus weighting 3.6 kg with an id of 3.1 cm and diameter matches that of the cylinder. Calculate its moment of inertia.

Solution: Using Eq:4.6 where $B = 0.059$ m, $A = 3.1$, $X = 3.6$ kg and $Y = 5.3$ kg, the *fx-991ES* keystrokes are,

1 ▦ 2 ▶ (([ALPHA] (B) x^2 (([ALPHA] (Y) + [ALPHA] (X))) +
[RCL] (X) [ALPHA] (A) x^2) [CALC]

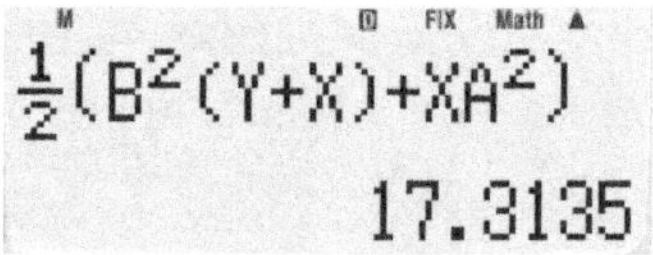

M   D   FIX   Math ▲

$\frac{1}{2}(B^2(Y+X)+XA^2)$

17.3135

The moment of inertia of the compound object is 17.3135 kg.m^2.

---

5. Rotation

Everyone will have recognised that mechanical system involve rotation in one form or another. You are already familiar with linear velocity and when describing rotating objects the phrase angular velocity is often used. In the context of machines this is measured in *revolutions per minute* or RPM. Many cars have a *rev counter* on the dashboard as shown by the image on the right as this shows the rotation rate of the engine. Typically, in a car travelling at 70 miles per hour, the engine

would be rotating at 2,000 RPM. The higher the rotation rate of the engine the more power the engine is producing.

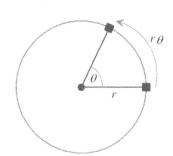

Consider the diagram on the left, the object is rotating in a circle of radius r. The angular velocity is given by,

$$\omega = \frac{d\theta}{dt}$$

This is the rate at which the angle θ is increasing. It's customary to use radians when dealing with angular velocity (rads/second). Referring to the diagram, the distance moved by the object as it

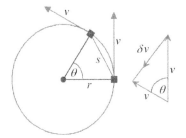

sweeps out an angle θ is $r\theta$, the linear velocity v is therefore,

$$v = \frac{d}{dt}(r\theta) = r\frac{d\theta}{dt} = r\omega$$

5.1

The period of revolution is the time taken for the object to complete one revolution. You will note as the object rotates its linear velocity is changing continuously as shown by the diagram on the left. By virtue of the changing velocity, an acceleration is involved in rotating objects. From the diagram there are two similar triangles,

$$\frac{s}{r} = \frac{\delta v}{v}$$

s is the distance travelled by the object in time δt, therefore $s = v\ \delta t$. Dividing these,

$$\frac{s}{r}\frac{1}{s} = \frac{\delta v}{v}\frac{1}{v\ \delta t}$$

The acceleration along the radius of the circle is therefore,

$$a = \frac{\delta v}{\delta t} = \frac{v^2}{r} \qquad\qquad 5.2$$

From Eq:5.1, $v = r\ \omega$, the acceleration is also expressed as,

$$a = r\omega^2 \qquad\qquad 5.3$$

Example 1: A particle of mass 2.1 kg is moving in a horizontal circle and the force acting upon it is 94.5 N directed towards the centre. If the radius of the circle is 5.1 m, what is it angular velocity?

Solution: Using Eq:5.3, force F = mass × acceleration, therefore,

$$F = rm\omega^2 \text{ or } \omega = \sqrt{\frac{F}{rm}}$$

The keystrokes for your *fx* calculator are, for F = 94.5 N, D = 5.1 m and M = 2.1 kg

$\boxed{\sqrt{\blacksquare}}$ $\boxed{\text{▤}}$ $\boxed{\text{ALPHA}}$ (F) $\boxed{\blacktriangledown}$ $\boxed{\text{ALPHA}}$ (D) $\boxed{\text{ALPHA}}$ (M) $\boxed{\text{CALC}}$

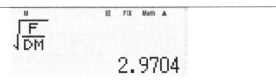

$$\sqrt{\frac{F}{DM}}$$

| M | ⊞ FIX Math ▲ |

$$2.9704$$

The angular velocity is 2.9704 rads.s^{-1}.

Example 2: A particle of mass 2.6 kg is travelling in a horizontal circle whose radius is 4.3 m. A force of 12.5 N is acting upon it, calculate the time it takes to complete one revolution.

Solution: Using Eq:5.3 and taking into consideration the force F,

$$F = rm\omega^2 \text{ or } \omega = \sqrt{\frac{F}{rm}}.$$

If ω is the angular velocity, the period of rotation T will be,

$$T = \frac{2\pi}{\omega} = 2\pi\sqrt{\frac{rm}{F}}$$

The keystrokes for your *fx* calculator with F = 12.5 N, M = 2 kg and D = 4 m are,

[2] [SHIFT] (π) [√■] [▤] [ALPHA] (D) [ALPHA] (M) [▼] [ALPHA] (F) [CALC]

| M | ⊞ FIX Math ▲ |

$$2\pi\sqrt{\frac{DM}{F}}$$

$$5.9422$$

The period of rotation will be 5.94 seconds.

In the last chapter there was a discussion on the moment of inertia, when an object with a moment of inertia I is rotating with an angular velocity ω, its kinetic energy is given by,

$$KE = \frac{1}{2}I\omega^2 \qquad\qquad 5.4$$

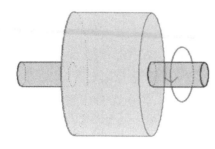

It is not uncommon to store energy in large rotating flywheels as shown in the diagram on the left. The energy is stored in the large section and the shaft reside within highly lubricated bearings so there is a minimum energy loss through friction. The moment of inertia for a solid cylinder is given in Eq:4.4, the kinetic energy stored by the flywheel is therefore,

$$KE = \frac{1}{2}\left(\frac{1}{2}mr^2\right)\omega^2 = \frac{m}{4}(r\omega)^2 \qquad 5.5$$

where r is the radius of the flywheel.

5.1 Rotating Flywheel

A typical problem involves a flywheel with a weightless rope with a weight attached to its end as illustrated in the figure on the right. Once the weight is released, it will move downwards causing the flywheel to turn. Both flywheel and weight will continue to accelerate until the weight reaches the floor. The parameters include,

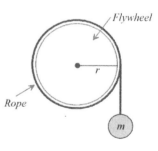

- Potential and kinetic energy of the both weight and flywheel.
- Rotation rate after t seconds.
- Distance d travelled by the weight.
- Tension in the rope.

The weight has potential energy and as it falls, this will be converted into rotational kinetic energy of the flywheel and kinetic energy of the descending weight. When considering a rotating flywheel that is accelerating, it's customary to define the *rotational motion L* by,

$$L = I\dot{\omega} \qquad\qquad 5.6$$

where I is the moment of inertia and $\dot{\omega}$ is the angular acceleration.

Example 3: A flywheel of mass 2 kg has a rope wrapped around it with a weight of 0.5 kg attached to the end. The diameter of the flywheel is 0.25 m. Calculate the following after the weight reaches 2 ms^{-1}:
1. The kinetic energy of the wheel.
2. The kinetic energy of the weight.
3. The distance travelled by the weight.

Solution: When the weight reached a velocity of 2 ms^{-1}, the velocity on the circumference of the flywheel will be the same, therefore
$v = r\omega$ giving an angular velocity of,
$$\omega = \frac{v}{r}. \qquad\qquad a$$
The moment of inertia of the wheel, this will be
$$I = \frac{1}{2}mr^2$$
The kinetic energy of the wheel is,
$$KE_{wheel} = \frac{1}{2}I\omega^2 = \frac{1}{4}mr^2\omega^2 \qquad\qquad b$$

The kinetic energy of the weight is
$$KE_{weight} = \frac{1}{2}mv^2 \qquad\qquad c$$
The potential energy lost to the weight is the sum of the kinetic energy of the flywheel and moving weight,

---

64

$$mgd = KE_{wheel} + KE_{weight}$$

The distance travelled d is,
$$d = \frac{KE_{wheel} + KE_{weight}}{mg}$$

d

Let $X = KE_{wheel}$, $Y = KE_{weight}$, $M = 2$kg, $A = 2$ ms^{-1}, $C = 0.25$ m, $E = 0.5$ mg, B is the angular velocity and D is distance travelled. The fx keystrokes for calculating Eq:a to Eq:d are,

a [ALPHA] (B) [ALPHA] (=) [ALPHA] (A) [▤] [ALPHA] (C) [▶] [ALPHA] (:)

b
[ALPHA] (X) [ALPHA] (=) [1] [▤] [4] [▶] [ALPHA] (M) [RCL] (C) [x^2] [RCL] (B) [x^2] [ALPHA] (:)

c [ALPHA] (Y) [ALPHA] (=) [1] [▤] [2] [▶] [ALPHA] (E) [RCL] (A) [x^2] [ALPHA] (:)

d [ALPHA] (D) [ALPHA] (=) [▤] [RCL] (X) [+] [RCL] (Y)
[▼] [RCL] (E) [SHIFT] (CONST) [3] [5] [CALC]

M		⊡ FIX Math ▲Disp
$B = \frac{A}{C}$		
		8.0000

M		⊡ FIX Math ▲Disp
$X = \frac{1}{4}MC^2B^2$		
		2.0000

M		⊡ FIX Math ▲Disp
$Y = \frac{1}{2}EA^2$		
		1.0000

M		⊡ FIX Math ▲
$D = \frac{X+Y}{E9}$		
		0.6118

The angular velocity is 8 rads.s^{-1}, the kinetic energy of the flywheel is 2 J, the kinetic energy of the weight is 1 J and the distance travelled by the weight is 0.6118 m.

3-1: Using the same keystrokes as *Example 3* with $M = 4.1$ kg, $A = 1.9$ ms^{-1}, $C = 0.32$ m, $E = 0.64$ mg, calculate the kinetic energy of the flywheel, the kinetic energy of the weight and the distance travelled by the weight.

Ans: 5.9 rads.s⁻¹, 3.7 J, 1.155 J, 0.77 m

Example 4: A weight of 2 kg is attached to a rope that is wrapped around a flywheel of radius 2 m and mass 5 kg. The weight is 3 m above the ground and is released. On reaching the calculate,

    1. The angular velocity of the wheel.
    2. The time taken for the weight to reach the ground.

Solution: The flywheel will accelerate; just before the weight hits the ground, all its potential will have been converted into the kinetic energy of the flywheel and the weight itself. If it's fallen through a distance h, the

$$\frac{1}{2}mv^2 + \frac{1}{2}I\omega^2 = mgh$$

The linear velocity v is related to the angular velocity by Eq:5.1, therefore,

$$\frac{1}{2}m(r\omega)^2 + \frac{1}{2}I\omega^2 = mgh$$

or

$$\omega^2(I + mr^2) = 2mgh$$

The angular velocity is therefore,

$$\omega = \sqrt{\frac{2mgh}{I+mr^2}} \qquad\qquad a$$

First calculate the moment of inertia for the flywheel,

$$I = \frac{1}{2}Mr^2 \qquad\qquad b$$

I is calculated first and the [Ans] is then put into Eq:a. Let X = 2 kg (attached weight), Y = 5 kg (mass of flywheel), D = 2 m

66

(radius) and A = 3 m (height about the ground). The keystrokes for your *fx* calculator are,

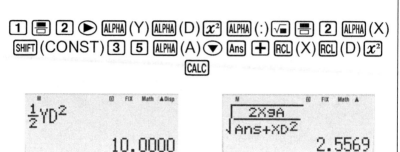

$$\frac{1}{2}YD^2$$

10.0000

$$\sqrt{\frac{2X9A}{Ans+XD^2}}$$

2.5569

The moment of inertia is 10 kg.m² and the angular velocity is 2.55 rads.s⁻¹.

4-1: Using the same keystrokes as in *Example 4*, the flywheel has a mass of 3.2 kg and a radius of 2.3 m. The end of the sting, which is wrapped around the flywheel, is attached to a 4 kg weight which is 2.2 m above the ground. After it has been released, calculate the angular velocity of a flywheel as the weight touches the ground.

Ans: 2.414 rads.s⁻¹.

5.2 Conical Pendulum

A pendulum on the end of piece of string that is mapping out a circle is referred to a *rotating conical pendulum* as illustrated in the diagram on the right. Resolving the vertical forces, we have,

$$mg = T\sin(\theta)$$

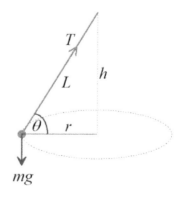

The tension in the string it therefore,

$$T = \frac{mg}{\sin(\theta)}$$
5.7

The horizontal forces are,

$$T\cos(\theta) = \frac{mv^2}{r}$$

Leaving,

$$T = \frac{mv^2}{r\cos(\theta)}$$
5.8

Combining Eq:5.7 and Eq:5.8,

$$v = \sqrt{\frac{gr}{\tan(\theta)}}$$
5.9

Using the identity, $1 + \tan^2(\theta) = \sec^2(\theta)$, and from the diagram, $\tan(\theta)$ can be expressed in terms of L and r to give,

$$v = r\sqrt{\frac{g}{\sqrt{L^2-r^2}}}$$
5.10

You will observe the velocity is not dependent on the mass of the pendulum. The period of revolution is the ratio of circumference to the velocity, therefore,

$$t = \frac{2\pi r}{v}$$

and using Eq:5.10,

$$t = 2\pi\sqrt{\frac{\sqrt{L^2-r^2}}{g}}$$
5.11

Example 4: A weight of 20 g is attached to a string of length 8 cm is circumscribing a circle of radius 5 cm. Calculate the tension in the string and the linear velocity.

68

Solution: Use Eq:5.7 to calculate the tension, replace $\sin(\theta)$ with $\sqrt{1 - \cos^2(\theta)}$ and since $\cos(\theta) = \dfrac{r}{L}$, to give,

$$T = \frac{mg}{\sqrt{1-\left(\frac{r}{L}\right)^2}}$$

Then use Eq:5.10 to calculate the linear velocity. Let M (mass) = 0.02 kg, A (pendulum length) = 0.08 cm and B (radius) = 0.05 m. The keystrokes for your *fx-991ES* calculator are,

[▤] [ALPHA] (M) [SHIFT] (CONST) [3] [5] [▽] [√▪] [1] [─] [(] [ALPHA] (B)
[▤] [ALPHA] (A) [▷] [)] [x²] [▷] [▷] [ALPHA] (:) [RCL] (B) [√▪] [SHIFT]
(CONST) [3] [5] [▤] [√▪] [RCL] (A) [x²] [─] [RCL] (B) [x²] [CALC]

The tension in the string is 0.251 N and the linear velocity is 0.626 ms^{-1}.

Example 5: A conical pendulum is that is made up of a non-extendable string and a weight. It is mapping out a circle of radius 30 cm with an angular speed of 4 rads.s^{-1}. Calculate the inclination angle from the horizontal.

Solution: Use Eq:5.1 and Eq:5.9 and rearrange to give with D the radius and X the speed,

$$\theta = \tan^{-1}\left(\frac{g}{DX^2}\right)$$

The *fx* keystrokes for this expression, when $D = 0.3$ m and , $X = 4$ rads.s^{-1},

[SHIFT] (tan^{-1}) [▤] [SHIFT] (CONST) [3] [5] [▽] [ALPHA] (D) [ALPHA] (X) [x²]
[▷] [)] [CALC]

$$\tan^{-1}\left(\frac{3}{D\times 2}\right)$$

$$63.91992587$$

The angle of elevation to the horizontal is 64°.

5.3 Vertical Rotating Objects

Having looked at objects rotating in the horizontal plane, we now focus our attention on an object rotating in the vertical plane as illustrated in the diagram on the right. As the object circumscribes a circle, its potential energy and kinetic energy are continuously changing. Also the tension T in the string will be changing. The forces acting on the object are given by,

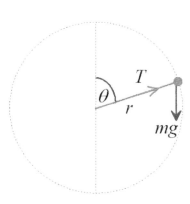

$$mg\cos(\theta) + T = \frac{mv^2}{r}$$

The tension is therefore given by,

$$T = m\left[\frac{v^2}{r} - g\cos(\theta)\right] \qquad\qquad 5.12$$

The sum of the potential and kinetic energies remain constant.

Example 6: A weight of 50 g is performing a vertical circular motion, and the radius of the circle is 2 m. The weight has a velocity of 5 ms⁻¹ when the string angle is 30° with the vertical. Calculate the tension in the string.

Solution: Use Eq:5.12 and let M = 0.05 kg, D (radius) = 2 m, X (velocity) = 5 ms⁻¹ and A (angle) = 30°. The fx keystrokes are,

$$\boxed{\text{ALPHA}\,(M)\,\text{(}\,\square\,\text{ALPHA}\,(X)\,\boxed{x^2}\,\blacktriangledown\,\text{ALPHA}\,(D)\,\blacktriangleright\,\boxminus\,\text{SHIFT}\,(\text{CONST})}$$

$$\boxed{3}\ \boxed{5}\ \boxed{\cos}\ \text{ALPHA}\,(A)\,\text{)}\,\text{)}\ \boxed{\text{CALC}}$$

M   D FIX Math ▲

$$M\left[\frac{X^2}{D}-g\cos(A)\right]$$

$$0.2004$$

The tension in the string is 0.2 N.

When a mass m is lying in an idle position, it can be subjected to a horizontal impulse I causing it to move in a circular pattern as illustrated in the diagram on the right. The angle through which the mass passes is θ and it raises to a height h_2 and a time of t seconds. You will observe that,

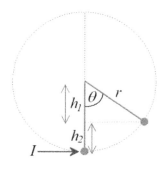

$$r = h_1 + h_2$$

Also,

$$r\cos(\theta) = h_2$$

Leaving,

$$h_2 = r - r\cos(\theta) = r[1 - \cos(\theta)]$$

The potential energy of the mass at height h_2 becomes,
$$mg[1 - \cos(\theta)]$$

If the velocity after the impulse is v_1 and it becomes v_2 after travelled through an angle θ then by the conservation of energy,

$$\frac{1}{2}mv_1^2 = \frac{1}{2}mv_2^2 + mgr[1 - \cos(\theta)]$$

This equation can be rearranged to give,

71

$$v_2 = \sqrt{v_1^2 - 2gr[1 - \cos(\theta)]} \qquad\qquad 5.13$$

You can see from this equation that v_2 will remain positive provided,

$$v_1^2 > 2gr[1 - \cos(\theta)]$$

This result can also be recognised from Eq:5.12, the tension T in the string will remain positive provided this condition is valid.

Example 7: A weight is lying on the end of a string of length 0.37 m in a vertical when it is subjected to an impulse giving it a velocity of 2.3 ms^{-1}. Calculate the velocity of the weight after it has passed through an angle of 41°.

Solution: The velocity can be calculated using Eq:5.13; this equation can be re-written using the following *fx-991ES* letter designations,

$$\sqrt{A^2 - 2gD[1 - \cos(E)]}$$

where A is the initial velocity, the string length is D and the angle is E. The *fx-991ES* keystrokes are,

[√■] [ALPHA] (A) [x^2] [—] [2] [SHIFT] (CONST) [3] [5] [ALPHA] (D) [(] [1]
[—] [cos]
[ALPHA] (E) [)] [)] [CALC]

A? [2] [·] [3] [=]
D? [0] [·] [3] [7] [=]
E? [4] [1] [=]

           D   Math ▲

$\sqrt{A^2 - 2gD(1 - \cos(E\blacktriangleright}$

1.873485202

The velocity after passing through an angle of 41° is 1.87 ms^{-1}.

7-1: Using the same keystrokes as *Example 7*, if the initial velocity is 1.9 ms^{-1} what is the velocity after the weight has passed through 35°?

Ans: 1.51 ms^{-1}.

Example 8: A weight whose mass is 0.3 kg is attached to a string of length 0.8 m. When hanging motionless the weight is given an initial velocity of 5 ms^{-1} which causes it to follow a partial vertical circular path. Calculate the following,
 1. The velocity of the weight when the string is horizontal.
 2. The tension in the string when it's horizontal.
 3. The angle with the vertical when the string slackens.

Solution: There is a transfer of energy from kinetic to potential, therefore
$$\frac{1}{2}mu^2 = \frac{1}{2}mv^2 + mgr$$
Rearranging this expression,
$$v = \sqrt{u^2 - 2gr} \qquad\qquad a$$
The tension in the string when it's horizontal is given by Eq:5.12 with $\cos(\theta) = 0$,
$$T = \frac{mv^2}{r} \qquad\qquad b$$
The string slackens when tension is zero, therefore from Eq:12,
$$\frac{V^2}{r} = g\cos(\theta)$$
where V is the velocity when this happens. By the conservation of energy,
$$\frac{1}{2}mu^2 = \frac{1}{2}mV^2 + mgr[1 + \cos(\theta)]$$
or

73

$$u^2 = V^2 + 2gr[1 + \cos(\theta)]$$

Substituting for V^2,

$$\theta = \cos^{-1}\left(\frac{u^2 - 2gr}{3gr}\right) \qquad\qquad c$$

To perform this calculation on your *fx* calculator, let A = 5 ms⁻¹ (initial velocity u), D = 0.8 m (radius r), B = v, M = 0.3 kg (particle mass m); the keystrokes are,

a

[ALPHA] (B) [ALPHA] (=) [√■] [ALPHA] (A) [x²] [−] [2] [SHIFT] (CONST) [3] [5]
　　[ALPHA] (D) [▶] [ALPHA] (:)
b [圖] [ALPHA] (M) [RCL] (B) [x²] [▼] [RCL] (D) [▶] [ALPHA] (:)
c
[SHIFT] (cos⁻¹) [圖] [RCL] (A) [x²] [−] [2] [SHIFT] (CONST) [3] [5] [RCL] (
D)
　　[▼] [3] [SHIFT] (CONST) [3] [5] [RCL] (D) [▶] [ )] [CALC]

$B = \sqrt{A^2 - 2gD}$	$\frac{MB^2}{D}$	$\cos^{-1}\left(\frac{A^2 - 2gD}{3gD}\right)$
3.0511	3.4910	66.7005

The velocity when the particle is horizontal is 3.05 ms⁻¹, the tension in the string is 3.49 N and the angle at which the string slackens is 66.7°.

8-1: Using the same keystrokes as *Example 7*, a weight whose mass is 0.55 kg is attached to a string of length 0.97 m. It is given an initial velocity of 6.3 ms⁻¹. Calculate the following, the velocity of the weight when the string is horizontal, the tension in the string when it's horizontal and the angle with the vertical when the string slackens.

Ans: 4.54 ms⁻¹, 11.71 N, 43.6°

6. Harmonic Motion

Oscillations are found almost every mechanical structure. Some are desirable, such as a watch mechanism, others are not, the unwanted vibrations in loudspeaker cabinets. Although the science of vibrations is quite complex we can start by looking at the most simplest vibration known as simple harmonic motion (SHM). Consider a vibration modelled by a sine wave,

$$x = A_o \sin(\omega t) \qquad\qquad 6.1$$

A_o is the peak amplitude of the sine wave and ω is its angular frequency measured in rads.s^{-1}. Angular frequency is related to normal frequency by the expression,

$$\omega = 2\pi f = \frac{2\pi}{T} \qquad\qquad 6.2$$

Below is a plot of the sine wave in Eq:6.1.

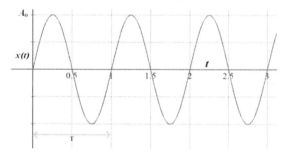

You will note the cyclic motion of the function and you can also see how the period T is defined, the time taken for the waveform to complete one cycle before it repeats. You will note in this example the period is 1 second the frequency is also 1 Hz. Now differentiate Eq:6.1,

$$\frac{dx}{dt} = \omega A_o \cos(\omega t)$$

Differentiate this expression,

$$\frac{d^2x}{dt^2} = -\omega^2 A_o \sin(\omega t)$$

Now substitute Eq:6.1 into this equation, giving

$$\frac{d^2x}{dt^2} = -\omega^2 x$$

or

$$\frac{d^2x}{dt^2} + \omega^2 x = 0 \qquad\qquad 6.3$$

This is the equation for a simple harmonic oscillator, it is a *second order differential equation* where ω is the angular frequency of the oscillator. This can be used to model many types of vibrations.

6.1 A Simple Pendulum

The figure on the right shows a simple pendulum swinging along an arc of a circle whose centre is at point O. You will be familiar with the nature of this oscillation. When displaced and released from point A its swings to point B and then comes to

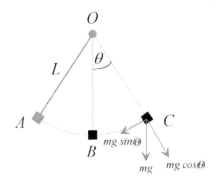

a halt at point C and then retraces it path. At points A and C the pendulum is stationary and at point B it has its greatest velocity. The maximum angle θ is shown in the diagram. The pendulum therefore swings backwards and forwards around the point O. When at point A or C the restoring force of gravity and its component along the tangent of the circular path is *mg sin*(θ). The *moment* about the point O is *mgL sin*(θ) together with the rotational moment (Eq:5.6) $I\ddot{\omega}$ acting in the same direction. Therefore,

$$mgL \sin(\theta) = -mL^2 \frac{d^2\theta}{dt^2}$$

76

This can be written as,

$$\frac{d^2\theta}{dt^2} + \frac{g}{L}\sin(\theta) = 0$$

When the angle is small then $\sin(\theta) \approx \theta$, therefore

$$\frac{d^2\theta}{dt^2} + \frac{g}{L}\theta = 0 \qquad\qquad 6.4$$

Comparing Eq:6.4 with Eq:6.3, the angular frequency of the pendulum is,

$$\omega = \sqrt{\frac{g}{L}} \qquad\qquad 6.5$$

Using Eq:6.2, the period of oscillation is given by,

$$T = 2\pi\sqrt{\frac{L}{g}} \qquad\qquad 6.6$$

The oscillation frequency is not dependent upon the mass only the length of the pendulum string. Probably you will already have seen a clock with a pendulum as shown in the image on the left. Clearly visible is the weight which is suspended from the clock mechanism behind the clocks' face. You will notice the pendulum rod extends beyond the base of the weight. The pendulum position on the rod can be changed. This effectively allows the value of L to be varied to enable corrections to be made to the value of T. This is one application where the simple pendulum has an application. The

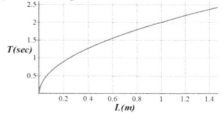

energy in the clock in lost in three places, the internal mechanism of the cog wheels, the pivot which is not frictionless and, to a lesser extent, air resistance. A plot of Eq:6.6 is shown on the right. This is a square root function which shows that if the pendulum length is 1 m the period of the swing is 2 seconds. Whereas a 1second oscillation requires a pendulum length of just over 20 cm.

6.2 Compound Pendulum

You may have come across a pendulum that comprises a distribution of mass throughout its object. Consider the object shown on the right of weight m. The centre of mass is at point C and it's supported at point P – the Pivot point. As you can imagine when it's displace and released it will oscillate around the point P. Since the mass of

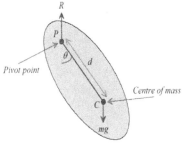

the object is distributed this is an example of compound pendulum. The anticlockwise force is $I\ddot{\theta}$ and the clockwise force is $mgd \sin(\theta)$ – this is the moment of the weight about the axis. Since it oscillates then,

$$mgd \sin(\theta) = -I\ddot{\theta} \qquad 6.7$$

For a small angle $\sin(\theta) \approx \theta$ leaving the angular acceleration as,

$$\ddot{\theta} = -\frac{mgd}{I}\theta \qquad 6.8$$

When you compare this equation with Eq:6.3, the angular frequency is,

$$\omega = \sqrt{\frac{mgd}{I}} \qquad 6.9$$

78

and the period of oscillation is,

$$T = 2\pi \sqrt{\frac{I}{mgd}}$$ 6.10

Any object of area A will have a *radius of gyration* which is given by,

$$k = \sqrt{\frac{I}{A}}$$ 6.11

The moment of inertia can be expressed as,

$$I = md^2 + mk^2 = m(d^2 + k^2)$$

The period of the oscillation from Eq:6.10 becomes,

$$T = 2\pi \sqrt{\frac{k^2 + d^2}{gd}}$$ 6.12

The distance d is the separation of the centre of gravity and the pivotal point and k is the radius of gyration about a parallel axis passing through the centre of gravity.

When Eq:6.12 is compared to a the period of a simple pendulum Eq:6.6, then,

$$\frac{L}{g} = \frac{k^2 + d^2}{gd}$$

The equivalent length for a simple pendulum is given by,

$$L = d + \frac{k^2}{d}$$ 6.13

Example 1: A compound pendulum is made up of a uniform rod of length b of mass m_r and a circular disc of radius a and mass m_d as shown in the diagram on the right. It can rotate about the point A. The distance of the centre of the disc to the point A is x. Derive an expression for the moment of inertia and the period of oscillation.

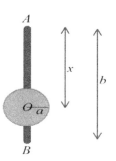

Solution: The moment of inertia of a rod about its horizontal axis through the mid-point is (Eq:4.2),

$$\frac{m_r b^2}{12}$$

The moment of inertia about the point A is,

$$\frac{m_r b^2}{12} + m_r \left(\frac{b}{2}\right)^2 = \frac{m_r b^2}{3}$$

The moment of inertia about a perpendicular axis of a disc through O is (Eq:4.3),

$$\frac{1}{2} m_d a^2$$

The moment of inertia of the disc about a parallel axis through A is,

$$\frac{1}{2} m_d a^2 + m_d x^2 = m_d \left(\frac{1}{2} a^2 + x^2\right)$$

The moment of inertia of the rod and the disc about A is therefore,

$$I = \frac{m_r b^2}{3} + m_d \left(\frac{1}{2} a^2 + x^2\right)$$

Using Eq:6.10, the period of oscillation is given by,

$$T = 2\pi \sqrt{\frac{I}{Mgh}}$$

80

In this expression h is the distance between the centre of gravity and the pivotal point. In this example the composite mass of the rod and the disc is $m_r + m_d = M$. The moments around the axis of rotation (point A) are,

$$m_r \frac{b}{2} + m_d x = Mh$$

Rearranging,

$$h = \frac{m_r + 2m_d}{2M} = \frac{bm_r + 2xm_d}{2(m_r + m_d)}$$

After a little algebra,

$$T = 2\pi \sqrt{\frac{m_d(a^2 + 2x^2)}{g(bm_r + 2xm_d)}}$$

Example 2: Using the two expression derived in *Example 1*, plot a graph of T against x with the following values: $m_r = 0.86$ kg, $b = 1.1$ m, $m_d = 1.3$ kg and $a = 0.11$ m. The x values should be separated by 10 cm.

Solution: Use the TABLE function in the *fx-991ES* calculator to derive the data for the graph. Set up the calculator in its table mode,

MODE 7 and enter the following keystrokes,

2 SHIFT (π) √■ ▤ 1 · 3 ((0 · 1 1 x^2 + 2
ALPHA (X) x^2) ▼ SHIFT (CONST) 3 5 (1 · 1 ×
0 · 8 6 + 2 RCL (X) × 1 · 3) =

Start? 0 · 1 =
End?   1 · 0 =
Step?   0 · 1 =

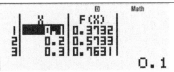

0. 1

Transfer the data values to a table where x is measured in cm and T is measured in seconds.

x	10	20	30	40	50	60	70	80	90	100
T	0.37	0.57	0.76	0.93	1.09	1.23	1.37	1.49	1.61	1.72

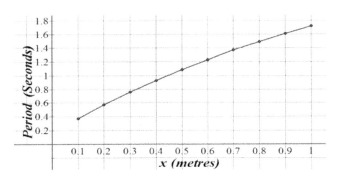

A plot of these data points is shown here. You will observe the curve is almost linear. As expected the further away the disc is from the pivotal point the longer the cycle time of the pendulum.

In *Example 2*, the calculations were performed on a specific compound pendulum. To generalise so that any rod length and disc diameter can be considered; calculations can performed on your *fx-991ES* calculator using the [CALC] key. Rod length is B (*b*), the disc radius is A (*a*), the rod mass is E (*m_r*) and the disc mass is F (*m_d*), the disc is a distance X (*x*) along the rod. The equation in *Example 1* for the period of oscillation of the compound pendulum can now be written as,

82

$$T = 2\pi \sqrt{\frac{F(A^2 + 2X^2)}{g(BE + 2XF)}}$$ 6.14

The keystrokes are,

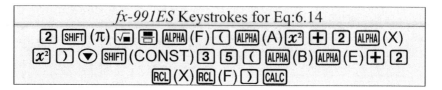

fx-991ES Keystrokes for Eq:6.14
[2] [SHIFT] (π) [√■] [🔲] [ALPHA] (F) [(] [ALPHA] (A) [x²] [+] [2] [ALPHA] (X)
[x²] [)] [▼] [SHIFT] (CONST) [3] [5] [(] [ALPHA] (B) [ALPHA] (E) [+] [2]
[RCL] (X) [RCL] (F) [)] [CALC]

Example 3: A compound pendulum comprises a rod of length 2.1 m and weight 1.7 kg and a disc of radius 5.4 cm and weight 0.35 kg. If the disc is 70 cm from the pivotal point, what is the period of oscillation?

Solution: Use Eq:6.14 *fx-991ES* keystrokes with the following values, B = 2.1 m, E = 1.7 kg, A = 0.054 m, F = 0.35 and X = 0.7 m,

$$2\pi \sqrt{\frac{F(A^2 + 2X^2)}{g(BE + 2XF)}}$$
$$0.5840482652$$

The period of oscillation is 0.58 seconds.

If you had piano lessons it's highly likely your piano teacher would have used a *metronome* as shown on the left. In principle it is an inverted compound pendulum where the position of the weight, as seen on the metal bar, can be moved along the bar. The period of oscillation (the tempo) can be varied from slow when the weight it at the top of the bar to fast when the weight is on the base of the bar. Most musical scores will have a metronome marking on them.

6.3 Weight on a Spring

You will probably have come across a weight attached to a spring, similar to the diagram shown on the right. When the weight is displaced and released it will oscillate. This is an example of simple harmonic motion. The frequency of its oscillation will be dependent upon the spring constant k which is related to the stiffness of the spring material. According to Hook's Law, if a spring is displaced by x, the restoring force F is proportional to the amount of displacement,

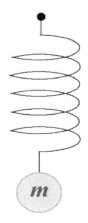

$$F = -kx$$

Since force = mass × acceleration then,

$$m\frac{d^2x}{dt^2} = -kx$$

This expression can be written as,

$$\frac{d^2x}{dt^2} + \frac{k}{m}x = 0 \qquad\qquad 6.14$$

When this equation is compared to Eq:6.3, the frequency of oscillation of the mass attached to the spring is,

$$\omega = \sqrt{\frac{k}{m}} \qquad\qquad 6.15$$

The oscillation frequency is dependent upon the weight of the mass and the spring constant k. The units for k are Nm^{-1}. The smaller the value of k the higher the natural oscillation frequency. The acceleration can be expressed as,

$$\frac{d^2x}{dt^2} = \frac{dv}{dt} = \frac{dv}{dx}\frac{dx}{dt} = v\frac{dv}{dx}$$

Therefore Eq:6.14 becomes,

$$v\frac{dv}{dx} + \frac{k}{m}x = 0 \qquad\qquad 6.16$$

The velocity at a distance a from its rest position is determined by integrating this expression,

$$\int v \, dv = -\frac{k}{m} \int_a^x x \, dx$$

This becomes,

$$\frac{v^2}{2} = \frac{k}{m} \int_x^a x \, dx$$

giving,

$$v = \sqrt{\frac{2k}{m} \int_x^a x \, dx} \qquad\qquad 6.17$$

where a is the amplitude. Alternatively performing the integration,

$$\frac{v^2}{2} = \frac{k}{m} \left[\frac{x^2}{2}\right]_x^a = \frac{k}{2m}[a^2 - x^2]$$

leaving,

$$v^2 = \frac{k}{m}[a^2 - x^2] \qquad\qquad 6.18$$

Using Eq:6.15,

$$v^2 = \omega^2[a^2 - x^2] \qquad\qquad 6.19$$

Example 4: A particle is oscillating according to Eq:6.14, its acceleration is 2.4 ms^{-2} when it is displaced 60 cm from its rest position. If the amplitude is 1.5 m, i. Calculate the speed when it's displaced 1.5 m from its rest position. ii. Calculate the distance from its rest position when the velocity is 2.4 ms^{-1}. iii. What is the maximum speed of the particle? iv. What is its maximum acceleration?

Solution: We need to calculate the ratio of the spring stiffness k to the mass m of the particle. The acceleration is given by

$$\frac{d^2x}{dt^2} = -\frac{k}{m}x$$

The value of $\frac{k}{m}$ can be calculated from,

$$\frac{1}{x}\frac{d^2x}{dt^2} = -\frac{k}{m} = \frac{1}{0.6} \times 2.4 = -\frac{k}{m} = -4, \ \frac{k}{m} = 4$$

i. The speed calculation, use Eq:6.18

$$v = \sqrt{\frac{k}{m}[a^2-x^2]} = \sqrt{4[1.5^2 - 1.2^2]} = 1.8 \ \text{ms}^{-1}$$

You can confirm this result on your *fx-991ES* with the following keystrokes for Eq:6.17,

$\boxed{\sqrt{\square}}\ \boxed{8}\ \boxed{\int_\square^\square}\ \boxed{\text{ALPHA}}\ (\text{X})\ \boxed{\blacktriangledown}\ \boxed{1}\ \boxed{\cdot}\ \boxed{2}\ \boxed{\blacktriangle}\ \boxed{1}\ \boxed{\cdot}\ \boxed{5}\ \boxed{=}\ \boxed{\text{S⇔D}}$

$$\sqrt{8\int_{1.2}^{1.5}Xdx}$$
$$1.8$$

ii. Rearrange Eq:6.18 to make x the subject,

$$x = \sqrt{a^2 - \frac{m}{k}v^2} = \sqrt{\left(1.5^2 - \frac{1}{4} \times 2.4^2\right)} = \pm 0.9 \ \text{m}$$

iii. the maximum occurs when the displacement x is zero, from Eq:1.8,

$$v_{max} = \sqrt{\frac{ka^2}{m}} = \sqrt{4 \times 1.5^2} = 3\text{ms}^{-1}$$

1: A mass is oscillating on a spring and the spring stiffness to mass ratio is 2.73. When the amplitude is 25 cm and the mass is 23 cm from its rest position, calculate,

1-i. its acceleration
1-ii. its velocity
1-iii the frequency of oscillation
1-iv the period of oscillation

Ans: 0.628 ms^{-2}, 0.162 ms^{-1}, 0.26 Hz, 3.8 s

2: A mass is oscillating on a spring and the spring stiffness to mass ratio is 7.48. When the amplitude is 15 cm and the mass is 9 cm from its rest position, calculate,

1-i. its acceleration
1-ii. its velocity
1-iii the frequency of oscillation
1-iv the period of oscillation

Ans: 0.67 ms^{-2}, 0.328 ms^{-1}, 0.435 Hz, 2.3 s

6.4 Sea Tides

An interesting example of SHM is the movement of tides. The plot below shows a simplified harbour tide height against hours.

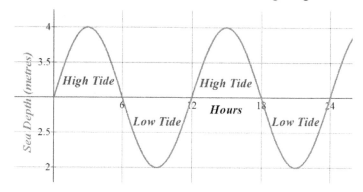

You may consider a harbour where the tide ebbs and flows twice every 24 hours. The sea depth rises and falls within the harbour. Harbours are usually designed to be navigated by ships at any time which means that even at low tide there should still be sufficient depth of water to accommodate most custom ships. The following example illustrates the case where this not true.

Example 5: On 26th June a harbour has a low tide of 5 m at 12:00. At the following high tide at 18:15 the tide height is 15 m. A particular ship requires a height of 12 m of water to navigate safely. Determine the earliest time during the day the ship can safely enter and leave the harbour.

Solution: The amplitude of the tide is the difference between the highest and lowest levels, this is $15 - 5 = 10$ m $= a$. The half period of the tide is $18:15 - 12:00 = 6.25$ hours, giving $T = 12.5$ hours. Therefore,

$$\frac{2\pi}{\omega} = 12.5 \quad \text{or} \quad \omega = \frac{4\pi}{25}$$

The height of the water y at time t is given by,

$$y = 15 + 10\sin(\omega t + b)$$

b is a phase value that is included to ensure the tide times are correct. At low tide we know that,

$$\sin\left(\frac{4\pi}{25}12 + b\right) = -1$$

Therefore

$$\frac{4\pi}{25}12 + b = \frac{3\pi}{2}$$

Leaving $b = -1.3195$ rads. The sea water height y at time t is,

$$y = 10 + 5\sin\left(\frac{4\pi}{25}t - 1.3195\right)$$

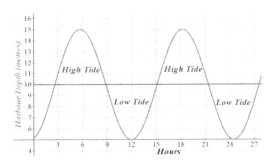

A plot of this function is shown on the previous page showing when the high and low tides occur. You can use the SOLVE feature on your *fx-991ES* to determine the times for a given sea water height. We need to know the times for a height $y = 12$ m.

When using SOLVE, need to enter guesses, if there is a low tide at 12:00, try 14:00 and also 21:00. Rewrite the equation as
$$y = 10 + 5\sin(0.5026t - 1.3195)$$

And use the following *fx-991ES* keystrokes (*X* is used instead of *t*),

[1] [2] [ALPHA] [(=)] [1] [0] [+] [5] [sin] [0] [.] [5] [0] [2] [6]
[ALPHA] [(X)] [−] [1] [.] [3] [1] [9] [5] [)] [SHIFT] (SOLVE)

Solve for X   [1] [4] [=]   Solve for X   [2] [1] [=]

```
 ⊞        Math ▲              ⊞        Math ▲
12=10+5sin(0.50▶         12=10+5sin(0.50▶
X=   15.94548777         X=   20.55861742
L-R=           0         L-R=           0
```

Note how you calculate the minute measurement from the result of the calculation, enter the following keystrokes,

[AC] [Ans] [° ' ''] [=]
```
            ⊞ FIX  Math ▲
Ans°

      15°56'43.76"
```
The sea height will be at 12 m at 15:56 and also at 20:33. The ship needs to enter the harbour after 15:56 in the afternoon and leave before 20:33 in the evening.

5-1: Use the same keystrokes as *Example 5*, by using the plot for guess values, determine the times when the tide was at 9 m.

Ans: 2:19, 9:14, 14:46, 21:47

6.5 Damped Harmonic Motion
When a weight is attached to a spring and is set into motion, it oscillates and the amplitude of the oscillations diminishes until it

eventually stops. The oscillations are damped owing to the energy loss in the spring. An equation that is used to model damping is,

$$y = e^{-\alpha t} \cos(\omega t) \qquad\qquad 6.20$$

The decay of the oscillation is modelled as an exponential decay where α is the damping factor. A plot of this function is shown on the right together with the exponential decay envelope. The greater the value of α the greater the rate of decay. This type of oscillation is called a *resonance* and the majority of sounds we hear are damped

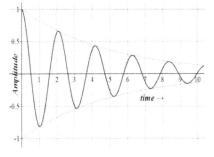

resonance sounds. The plot below is a one second recording from a piano.

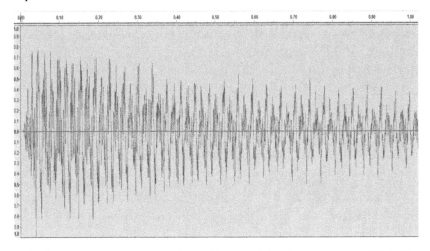

It is the lowest note on a piano (A0) and its frequency is 27.5 Hz. You will observe how the waveform is decaying, the complex nature of the waveform – there are many harmonics present in the sound. This is what gives it the distinctive piano sound. Below is a plot of damped oscillations with various degrees of damping.

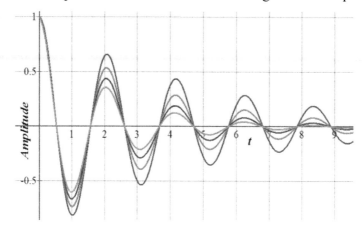

In this plot the value of α varies from 0.2 to 0.5 and you will observe that after 8 seconds one oscillation has almost stopped. All tones generated by musical instruments are damped oscillations. A measure of the damping is the time it takes for the amplitude of an oscillation to reduce to 50%. Consider the damping at two times,

$$y_1 = e^{-\alpha t_1} \text{ and } y_2 = e^{-\alpha t_2}$$

Need to determine the value of t_2 when $y_2 = 0.5y_1$. Therefore,

$$e^{-\alpha t_2} = 0.5 \, e^{-\alpha t_1}$$

For convenience let $t_1 = 0$ then in general

$$2 = e^{\alpha t}$$

The oscillation will have reduced to 50% when

$$t = \frac{\ln(2)}{\alpha} \qquad\qquad 6.21$$

Example 6: The following table contains data points for a damped harmonic oscillator. Determine the value of the damping factor and the frequency of the waveform.

t	0	0.5	1.0	1.5	2.0	2.5	3.0
y	0	0.06	-0.72	-0.13	0.5	0.15	-0.35
t	3.5	4.0	4.5	5.0	5.5	6.0	
y	-0.15	0.23	0.14	-0.15	-0.12	0.01	

Solution: Plotting these data points and joining the data values you will get a plot very similar to the one shown on the right. Mark on the plot the 50% lines (shown in red). The upper red line intersects the plot very close to time = 2 seconds. Using

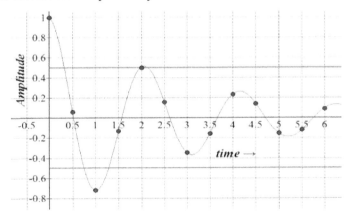

Eq:6.21 and making α the subject of the equation,

$$\alpha = \frac{\ln(2)}{2} = 0.346$$

The end of the second cycle is at 4.2 seconds, the frequency is therefore,

$$f = \frac{number\ of\ cycles}{time\ period} = \frac{2}{4.2} = 0.476\ Hz$$

An interesting application of damping is on sprung loaded doors that open in both directions (made famous in Westerns as doors to bars). The amount of swing depends on the degree of damping in the springs.

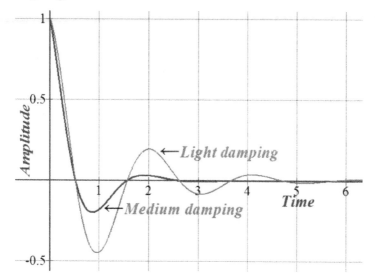

In this plot you can see two levels of damping; light damping where the doors swing several times before they finally come to rest and medium damping where after one swing they are nearly at rest. For a complete analysis of damping it's necessary to delve into the *second order differential equation* model for harmonic damping. You will then encounter, heavy damping, light damping and critical damping.

1: Use your *fx-991ES* calculator in its TABLE mode to calculate values for the function,
$$f(x) = e^{-0.8x} \cos(3x)$$
in the range $0 \le x \le 4$ in steps of 0.5 and place the data values on the graph on the next page.

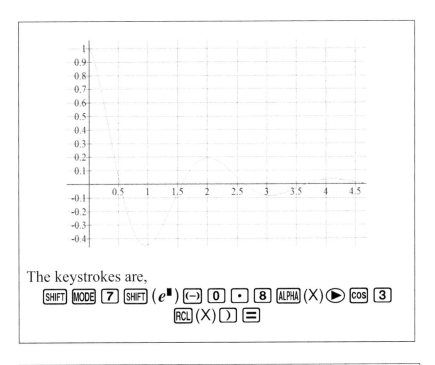

The keystrokes are,

SHIFT MODE 7 SHIFT ($e^{\blacksquare}$) (−) 0 · 8 ALPHA (X) ▶ cos 3
RCL (X)) =

2: The following table contains data points for a damped harmonic oscillator. Plot them out and determine the value of the damping factor and the frequency of the waveform.

t	0	0.25	0.5	0.75	1.0	1.25	1.5	1.75
y	1	0.51	-0.38	-0.85	-0.53	0.22	0.71	0.53
t	2.0	2.25	2.5	2.75	3.0	3.25	3.5	3.75
y	-0.1	-0.58	-0.51	0	0.46	0.47	0.07	-0.36

Ans: 0.22, 0.63 Hz

This completes the second book on mechanics from *Cambridge Paperbacks*. By now you should be an experienced user of the *fx-991ES* calculator which should have served you well in your understanding of mechanics.

Printed in Great Britain
by Amazon

87272504R00059